SCIENCE CHUNKS

Year C Guide

Science Chunks - Year C

First Edition 2023

ISBN: 978-1-953490-21-6

Email: support@elementalscience.com

Copyright Policy

Year C Overview

Welcome to Year C of the Science Chunks series! The seven units that make up this year will cover the living world, invertebrates, our planet, the periodic table, forces, and motion, plus a short unit on Marie Curie. You will learn about these subjects by:

- **Reading** the assigned encyclopedia pages with your students to learn about the lesson's topic.
- **Writing** down what the students have learned in a way that is appropriate for their skills using either the lapbook templates or the notebook sheets.
- **Doing** one or more of the lesson's coordinating STEAM activities.

What do we need?

In addition to the lesson plans found in the seven units included in the Year C bundle, you will need the following materials:

1. **Required Books** – These books are scheduled for the reading assignments:

   Younger Students (K to 3rd grade)
   - ✓ *Kingfisher My First Encyclopedia of Animals*, 2021 Edition
   - ✓ *Basher Science: The Complete Periodic Table*, 2015 Edition
   - ✓ *Basher Science Physics: Why Matter Matters!*, 2016 Edition
   - ✓ *The DK Children's Encyclopedia*, 2017 Edition

   Older Students (3rd to 6th grade)
   - ✓ *Animal! (DK and Smithsonian)*, 2016 Edition
   - ✓ *DK Eyewitness: The Elements*, 2022 Edition
   - ✓ *The Usborne Science Encyclopedia*, 2015 Edition
   - ✓ *The Kingfisher Science Encyclopedia*, 2017 Edition

   For the scientist study you will need a copy of *Who Was Marie Curie?* by Megan Stine. To see these books and get quick access to the links mentioned in the units, please visit the Year C resource page:

   https://elementalscience.com/blogs/resources/sc-c

2. **Supplies** – You will need various supplies for the lapbook or notebook, plus the coordinating activities. You can find a full list of what you need in the introductory material for each unit.

How do we organize the year?

Generally, we recommend completing one lesson per week, but please feel free to organize the units in the order that fits your goals. Here are two options:

Option 1 (By Discipline)
- ☐ Living World (4 Lessons)
- ☐ Invertebrates (5 Lessons)
- ☐ Planet Earth (5 Lessons)
- ☐ Periodic Table (12 Lessons)
- ☐ Curie (2 Lessons)
- ☐ Forces (5 Lessons)
- ☐ Motion (3 Lessons)

Option 2 (By Season - Starting in the Fall)
- ☐ Invertebrates (5 Lessons)
- ☐ Planet Earth (5 Lessons)
- ☐ Forces (5 Lessons)
- ☐ Curie (2 Lessons)
- ☐ Periodic Table (12 Lessons)
- ☐ Motion (3 Lessons)
- ☐ Living World (4 Lessons)

If you have any questions or comments as you work through these materials, please don't hesitate to let us know by emailing support@elementalscience.com.

Table of Contents

A Peek Inside a Science Chunks Unit

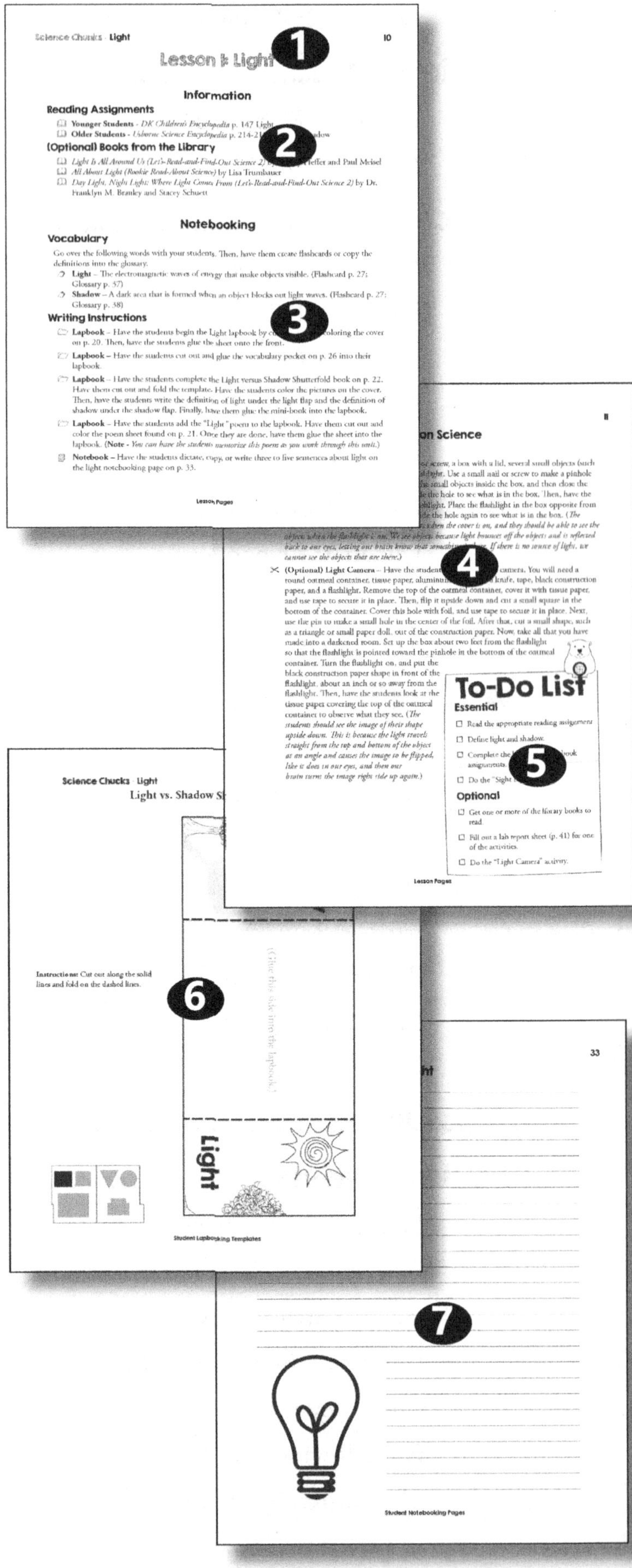

1. Lesson Topic

Focus on one main idea throughout the week. You will learn about these ideas by reading from visually appealing encyclopedias, recording what the students learned, and doing coordinating hands-on science activities.

2. Information Assignments

Find two reading options—one for younger students, one for older students, plus optional library books.

3. Notebooking Assignments

Record what your students have learned with either a lapbook or a notebook. The directions for these options are included for your convenience in this section along with the vocabulary the lesson will cover.

4. Hands-on Science Assignments

Get the directions for coordinating hands-on science activities that relate to the week's topic.

5. Lesson To-Do Lists

See what is essential for you to do each week and what is optional. You can check these off as you work through the lesson so that you will know when you are ready to move on to the next one.

6. Lapbook Templates

Get all the information you need to create a lapbook on the subject. *(Year bundle sold separately.)*

7. Notebook Templates

Have all the sheets you need to create a notebook on the subject, including a glossary for the vocabulary terms. *(Year bundle sold separately.)*

In the appendix you will find a blank activity sheet, a blank lab report sheet, and any necessary templates.

Introduction

Science Chunks - Year C is a unique and versatile unit study that leads you through a survey of the living world, invertebrates, planet Earth, the periodic table, forces, and motion, plus a short unit on Marie Curie. It is designed to be a gentle approach to homeschool science based on the Unit Study method suggested in *Success in Science: A Manual for Excellence in Science Education* by Bradley and Paige Hudson. This study can be used as a stand-alone unit for elementary science.

What Is Included in Each Unit

Science Chunks - Year C includes seven units, each with the three keys to teaching science. For a lesson you will be doing the following:

- ✓ Listening to (or reading) **scientific information** from visually appealing encyclopedias
- ✓ Dictating (or writing down) what the students have learned and seen using **lapbooking or notebooking**
- ✓ Watching (and doing) **hands-on science** through a variety of science activities

Here is how this works.

Section I - Information

The elementary student is an empty bucket waiting to be filled with information, and science-oriented books are a wonderful way to do that. These books can include age-appropriate children's science encyclopedias, living books for science, and/or children's nonfiction science books.

In this program, the reading assignments and additional books scheduled in the lesson fulfill this component. The reading assignments are broken for you into two levels: younger students (1st to 3rd grade) and older students (4th to 6th grade).

Our idea is that you will read these selections with your students, pausing to ask questions or discussing the information once you are done reading.

Section 2 - Notebooking

The purpose of the notebooking component for elementary science education is to verify that the students have placed at least one piece of information into their knowledge bucket. You can use notebooking sheets, lapbooks, and/or vocabulary words to fulfill this requirement.

In this program, we have created two writing options, a lapbook and a notebook, for you to use with your students. In *Science Chunks Year C Student Lapbook Templates*, you will find all of the templates and pictures you will need to complete seven different lapbooks for the year. In *Science Chunks Year C Student Notebook*, you will find all the pages you need to create a simple

SCIENCE CHUNKS

notebook documenting what the students learned this year, including notebooking sheets and a glossary for each unit.

Section 3 - Hands-on Science

Scientific demonstrations and observations are meant to spark students' enthusiasm for learning science, to work on their observation skills, and to demonstration the principles of science for them. This component of elementary science education can contain scientific demonstrations, hands-on projects, and/or nature studies.

In this program, the coordinating activities at the end of each lesson fulfill this section of elementary science instruction. If you would like to record what you have done, you can use one of the templates in the appendix pp. 102-103.

What You Need in Addition to This Guide

Books Scheduled

The following books are what we used to plan the reading assignments for this year:

- **Younger students (1st to 3rd grade)** - *Kingfisher My First Encyclopedia of Animals, Basher Science: The Complete Periodic Table, Basher Science Physics: Why Matter Matters!, The DK Children's Encyclopedia*
- **Older students (4th to 6th grade)** - *Animal! (DK and Smithsonian), DK Eyewitness: The Elements, Usborne Science Encyclopedia, Kingfisher Science Encyclopedia*
- **All students** - *Who Was Marie Curie?* by Megan Stine

However, you could certainly use the encyclopedias you already have on hand or books from the library.

Unlike the single Science Chunks units, this printed Year C guide does not inlcude the student materials. You will need to purchase one (or both) of the following:

- *Science Chunks Year C Lapbook Templates (LT)*
- *Science Chunks Year C Student Notebook (NB)*

In addtion to the books and the student materials, you will need also simple craft supplies and other science materials—see a complete list of essential items on pp. 11-12.

How This Unit Works

We have included a to-do list with each lesson to give you an idea of what is essential and what is optional. There are several ways you can schedule this unit. Here is a quick look at a few of the options.

Possible Schedules for Your Week

- **One Day** – You can set aside about an hour to an hour and a half each week to complete all the essential tasks in one day.

- **Two Days** – You can set aside about 30 to 40 minutes twice a week to complete all the essential tasks, plus a few more, in two days. On the first day, you can complete the reading assignments and either the lapbook or notebook assignments. On the second day, you can complete the coordinating activity and the vocabulary assignments as well as read any library books.
- **Three Days** – You can set aside about 30 minutes three times a week to complete all the essential tasks, plus a few more, in three days. On the first day, you can complete the reading assignments and either the lapbook or notebook assignments. On the second day, you can complete the coordinating activity and write a lab report using one of the templates. On the third day, you can do the vocabulary assignments as well as read any library books.
- **Four Days** – You can set aside about 20 to 30 minutes four times a week to complete all the essential tasks, plus a few more, in four days. On the first day, you can complete the reading assignments and either the lapbook or notebook assignments. On the second day, you can complete the coordinating activity and write a lab report. On the third day, you can do the vocabulary assignments as well as read any library books. On the fourth day, you can do the optional coordinating activity as well as read any library books.

Generally, we recommend completing one lesson per week, but please feel free to organize the units in the order that fits your goals.

Final Thoughts

Quick Links

View all the links mentioned in this guide in one place and get a digital copy of the appendix templates, glossary, and review sheets by visiting the following page:

https://elementalscience.com/blogs/resources/sc-c

Read Further

If you would like to read more about the philosophy behind the Science Chunks series, check out *Success in Science: A Manual for Excellence in Science Education* and the following articles from our website.

- **The Three Keys to Teaching Science** – This article shares the three keys to teaching science, including a free session that walks you through what each key can look like.

  https://elementalscience.com/blogs/news/3-keys
- **The Basics of Notebooking** – This article details the basic components of notebooking along with how a few suggestions on what notebooking can look like.

  https://elementalscience.com/blogs/news/what-is-notebooking
- **Lapbooking versus Notebooking** – This article takes a look at the differences between lapbooking and notebooking.

  https://elementalscience.com/blogs/news/lapbook-or-notebook

- **Scientific Demonstrations versus Experiments** – This article explains the difference between scientific demonstrations and experiments along with when and how to employ these methods.
    - https://elementalscience.com/blogs/news/89905795-scientific-demonstrations-or-experiments

Last Words

As the author and publisher of this curriculum, I encourage you to contact me with any questions or problems that you might have concerning *Science Chunks - Year C* by emailing us at support@elementalscience.com. I, or a member of our team, will be more than happy to answer them as soon as we can. I hope that you will enjoy creating memories this year!

~ Paige Hudson

Materials List

Living World Unit Materials

You will need the following materials to complete the essential coordinating activities:

- **Lesson 1:** At least 8 objects, some of which are living, some of which are not
- **Lesson 2:** Several friends, Several different colors of glitter, Tape
- **Lesson 3:** Gloves, Mushroom Cap, Paper, Glass Bowl, Water
- **Lesson 4:** Shoe box, Construction paper, Glue, Markers

Invertebrates Unit Materials

You will need the following materials to complete the essential coordinating activities:

- **Lesson 1:** Small, unglazed terracotta pot, Small rock or piece of mulch
- **Lesson 2:** Tweezers, Toothpick, Cotton Swab, Gummy Worm, Marshmallow, Lettuce Leaf
- **Lesson 3:** String (about 5 feet long)
- **Lesson 4:** Butterfly Life Cycle Pictures (Appendix p. 107)
- **Lesson 5:** Piece of fruit, Honey or syrup, Plate

Planet Earth Unit Materials

You will need the following materials to complete the essential coordinating activities:

- **Lesson 1:** Modeling (or air dry) clay (red, orange, yellow, blue, and green), ruler
- **Lesson 2:** Globe (or large ball), desk lamp, Post-it tab (or another type of removable marker)
- **Lesson 3:** Large bath towel, flat, smooth surface
- **Lesson 4:** Straight pin, milk jug, knife, magnet
- **Lesson 5:** Supplies will vary depending on your plan

Periodic Table Unit Materials

You will need the following materials to complete the essential coordinating activities:

- **Lesson 1:** LEGOs® in a variety of colors and sizes
- **Lesson 2:** Baking soda, vinegar, cup
- **Lesson 3:** Epsom salts, ammonia, water, a clear cup
- **Lesson 4:** Steel wool, vinegar, jar with lid
- **Lesson 5:** Plastic baggie, cup, gel glue, water, Borax
- **Lesson 6:** Cup, rock (limestone, if possible), vinegar
- **Lesson 7:** Can of dark cola soda, glass, several dirty pennies.
- **Lesson 8:** Candle, match, glass jar
- **Lesson 9:** 2 Eggs, 2 cups, fluoride toothpaste, plastic wrap, white vinegar
- **Lesson 10:** Helium-filled balloon, scissors

- ✂ **Lesson 11:** 3 Cups, 3 pencils, 3 different clear liquids (i.e., water, alcohol, and corn syrup)
- ✂ **Lesson 12:** Bite-sized candy or food (such as raisins or cereal puffs), timer
- ✂

Marie Curie Unit Materials

You will need the following materials to complete the essential coordinating activities:

- ✂ **Lesson 1:** White glue, water, iron filings, Borax, a few magnets
- ✂ **Lesson 2:** Cup, plate, food coloring, eye dropper, cookie cutter, baking soda, white vinegar

Forces Unit Materials

You will need the following materials to complete the essential coordinating activities:

- ✂ **Lesson 1:** Toy car, string, tape, several books, cardboard sheet
- ✂ **Lesson 2:** Bag of gumdrops, toothpicks, plate
- ✂ **Lesson 3:** Pencil, cardboard sheet (or hardcover book), flat surface
- ✂ **Lesson 4:** Marble, cookie sheet (or smooth cutting board), paper, felt, book
- ✂ **Lesson 5:** Balloon, old CD, 2-L soda bottlecap, thin nail, glue

Motion Unit Materials

You will need the following materials to complete the essential coordinating activities:

- ✂ **Lesson 1:** Straw, string (5 feet), scissors, large balloon, 2 chairs, tape
- ✂ **Lesson 2:** Toy car, ramp
- ✂ **Lesson 3:** Balloon, penny

Abbreviations Key
LT = *Science Chunks Year C Student Lapbook Templates*
NB = *Science Chunks Year C Student Notebook*

SCIENCE CHUNKS

Living World Unit

Lesson 1: The Living World

Information

Reading Assignments

- **All Students** - *"Is It Living or Nonliving?"* article on p. 42
- **Younger Students** - *DK Children's Encyclopedia* p. 38 Biology
- **Older Students** - *Kingfisher Science Encyclopedia* pp. 52-53 Classification of Living Things

(Optional) Books from the Library

- *Living Things and Nonliving Things: A Compare and Contrast Book* by Kevin Kurtz
- *What Is a Living Thing? (The Science of Living Things)* by Bobbie Kalman
- *What Do Living Things Need? (Science Readers)* by Elizabeth Austen

Notebooking

Vocabulary

Go over the following words with your students. Then, have them create flashcards or copy the definitions into the glossary.

- **Living** – Something that eats, breathes, and moves, like you. (Flashcard LT p. 15; Glossary NB p. 57)
- **Nonliving** – Something that is not alive. (Flashcard LT p. 16; Glossary NB p. 60)

Writing Instructions

- **Lapbook** – Have the students begin the Living World lapbook by cutting out and coloring the cover on LT p. 7. Then, have the students glue the sheet onto the front.
- **Lapbook** – Have the students complete the Living or Nonliving fold-book on LT p. 8. Have them cut out the pages and color the cover. Next, have the students tell you what they have learned about and label the pictures as living or nonliving. Then, have them fold the mini-book, glue on the cover, and add it to the lapbook.
- **Lapbook** – Have the students complete the Is It Living? flip-book. Have them cut out and color the mini-book found on LT p. 10. Then, have them finish the questions on the inside of the mini-book. Once they are done, have them glue the sheet into the lapbook.
- **Lapbook** – Have the students add the "Living Things" poem to the lapbook. Have them cut out and color the poem sheet found on NB p. 9. Once they are done, have them glue the sheet into the lapbook. (**Note** - You can have the students memorize this poem as you work through this unit.)
- **Notebook** – Have the students dictate, copy, or write three to five sentences about living and nonliving things on the Living World notebooking page on NB p. 6.

Hands-on Science

Coordinating Activities

✂ **Is it living?** – You will need at least eight objects, some of which are living, some of which are not. (If you cannot have actual living things, use pictures from books and magazines.) Set the objects in front of the students and have them examine each object as you help them answer the following questions with yes or no:

- **?** Does it change, develop, or grow?
- **?** Does it reproduce?
- **?** Does it need food?
- **?** Does it move?
- **?** Does it adapt or respond to what is around it?

You can have the students record what they have observed by writing the object name and the answers to the questions on the living or nonliving chart.

✂ **(Optional) Living and Nonliving** – Have the students play a game of Living and Nonliving. This is a simple game that is perfect for review. It can be downloaded for free from the You've Got This Math blog at the bottom of this post:

https://youvegotthismath.com/2016/08/16/livingvsnonlivingunit/

To-Do List

Essential

- ☐ Read the appropriate reading assignment.
- ☐ Define living and nonliving.
- ☐ Complete the lapbook or notebook assignments.
- ☐ Do the "Is is living?" activity.

Optional

- ☐ Get one or more of the library books to read.
- ☐ Fill out a lab report sheet (pp. 102-103) for one of the activities.
- ☐ Do the "Living and Nonliving" activity.

Lesson 2: Single-celled Organisms

Information

Reading Assignments

- **Younger Students** - *DK Children's Encyclopedia* p. 165 Microscopic Life
- **Older Students** - *Kingfisher Science Encyclopedia* p. 54 Single-celled Organisms

(Optional) Books from the Library

- *You Wouldn't Want to Live Without Bacteria!* by Roger Canavan
- *Protists: Algae, Amoebas, Plankton, and Other Protists* by Rona Arato
- *Bacteria: Staph, Strep, Clostridium, and Other Bacteria* by Judy Wearing

Notebooking

Vocabulary

Go over the following words with your students. Then, have them create flashcards or copy the definitions into the glossary.

- **Bacteria** – Single-celled organisms that live almost everywhere. (Flashcard LT p. 16; Glossary NB p. 52)
- **Protist** – A living thing whose body is made from a single cell. (Flashcard LT p. 17; Glossary NB p. 61)

Writing Instructions

- **Lapbook** – Have the students do the Single-celled Organisms flap-book on LT p. 11. Have them cut out the pages and color the pictures. Have them add information about bacteria, viruses, and protists on the respective pages. Then, have the students assemble the flap-book, staple the pages together, and add it to the lapbook.

- **Notebook** – Have the students dictate, copy, or write three to five sentences about bacteria, protists, and viruses on the Single-celled Organisms notebooking page on NB p. 7.

Hands-on Science

Coordinating Activities

- **Spreading Organisms** – You will need several friends, several different colors of glitter, and clear tape. Have each student (or participating person) choose a different color of glitter. Next, have them each rub some of the chosen glitter on their hands. Then, have everyone go around shaking each other's hands. After everyone has shaken each person's hands, have the students look at their hands and observe what has happened. Use a piece of clear tape to remove a sample of the glitter left on their hands, and stick the tape in the box on sheet of paper to observe.

- **(Optional) Virus Models** – Have the students make a model of several different types of viruses. You can find the directions for this project here:
  - Virus Models: https://www.science-sparks.com/easy-virus-models/

To-Do List

Essential

- ☐ Read the appropriate reading assignment.
- ☐ Define bacteria and protist.
- ☐ Complete the lapbook or notebook assignments.
- ☐ Do the "Spreading Organisms" activity.

Optional

- ☐ Get one or more of the library books to read.
- ☐ Fill out a lab report sheet for one of the activities.
- ☐ Do the "Virus Models" activity.

Lesson 3: Fungi and Mold

Information

Reading Assignments

- **All Students** - *"Mold"* article on p. 43
- **Younger Students** - *DK Children's Encyclopedia* p. 174 Mushrooms
- **Older Students** - *Kingfisher Science Encyclopedia* p. 55 Fungi and Lichens

(Optional) Books from the Library

- *Fungi (Kid's Guide to the Classification of Living Things)* by Elaine Pascoe, Janet Powell, and Dwight Kuhn
- *A Little Book of Slime: Everything That Oozes, from Killer Slime to Living Mold* by Clint Twist
- *Nature Close-Up—Slime, Mold and Fungi* by Elaine Pascoe

Notebooking

Vocabulary

Go over the following words with your students. Then, have them create flashcards or copy the definitions into the glossary.

- **Fungus** – A living thing that absorbs food from living or dead matter. (Flashcard LT p. 17; Glossary NB p. 55)
- **Spore** – A microscopic package of cells produced by a fungus or plant that can grow into a new individual. (Flashcard LT p. 18; Glossary NB p. 63)

Writing Instructions

- **Lapbook** – Have the students complete the Fungi and Mold wheel-book on LT p. 12. Have them cut out the pages and color the pictures. Have the students tell you what they have learned about fungi and molds and write their favorite piece of information next to the spaces on the wheel. Let the students decorate the cover. Once they are done, assemble the wheel-book with a brad fastener, and glue it into the lapbook.
- **Notebook** – Have the students dictate, copy, or write three to five sentences about fungus and mold on the Fungi and Molds notebooking page on NB p. 8.

Hands-on Science

Coordinating Activities

- **Spore Print** – You will need gloves, a mushroom cap, a sheet of paper, a glass bowl, and water. Get the mushroom cap ready for the students. You can use a button mushroom from your fridge by simply removing the mushroom stem, or you can harvest a cap from outside. If you choose to harvest an outdoor mushroom cap, put on the gloves, and go outside to cut off your selected mushroom cap. You can wrap it up in a flipped inside-out glove until the students are ready. Then, have the students set a piece of paper in a dark place where it won't be disturbed for the next 24 hours. Have them put on gloves and choose a mushroom cap. Have them observe the cap, noting the smooth side on top and the underside with the gills. Next, have the students place the cap, gill side down, on the center of the piece of paper. Sprinkle the cap with a bit of water, not too much, to moisten it a bit. Then, cover the cap with an inverted glass bowl. (**Next Day**) Remove the bowl, put the gloves back on, and gently peel back the mushroom cap to observe the spore prints that were made.

✂ **(Optional) Nature Walk** – Have the students take a walk in your local woods to look for examples of fungi and molds. (**Note** - Be sure to check fallen logs for signs of these living things.) You can have them observe the fruiting body with and without a magnifying glass. If you know that the fungus or mold is nontoxic, you can have them touch and smell it. NEVER touch a fungus that you are not familiar with.

To-Do List

Essential

- ☐ Read the appropriate reading assignment.
- ☐ Define fungus and spore.
- ☐ Complete the lapbook or notebook assignments.
- ☐ Do the "Spore Print" activity.

Optional

- ☐ Get one or more of the library books to read.
- ☐ Fill out a lab report sheet for one of the activities.
- ☐ Do the "Nature Walk" activity.

Lesson 4: Plants and Animals

Information

Reading Assignments

- **Younger Students** - *DK Children's Encyclopedia* p. 126 Habitats, p. 194 Plants, p. 22 Animal Groups
- **Older Students** - *Kingfisher Science Encyclopedia* pp. 68-69 Biomes and Habitats, pp. 56-57 Plant Anatomy, pp. 86-87 Mammals

(Optional) Books from the Library

- *Habitats (Science Readers: Content and Literacy)* by William B. Rice
- *Whose Habitat is That?* by Lucile Piketty and Lucy Brownridge
- *From Seed to Plant* by Gail Gibbons
- *A Tree Is a Plant (Let's-Read-and-Find-Out Science 1)* by Clyde Robert Bulla
- *The Fascinating Animal Book for Kids: 500 Wild Facts! (Fascinating Facts)* by Ginjer Clarke
- *Animal Sidekicks: Amazing Stories of Symbiosis in Animals and Plants* by Macken Murphy

Notebooking

Vocabulary

Go over the following word with your students. Then, have them create flashcards or copy the definitions into the glossary.

- **Habitat** – The place an animal or a plant calls home. (Flashcard LT p. 18; Glossary NB p. 55)

Writing Instructions

- **Lapbook** – Have the students do the Plants and Animals tab-book on LT pp. 13-14. Have them cut out the pages and color the pictures. Have them add information about habitats, plants, and animals on the respective pages. Then, have the students assemble the mini-book, staple the pages together, and add it to the lapbook.
- **Notebook** – Have the students dictate, copy, or write three to five sentences about habitats, plants, and animals on the Plants and Animals notebooking page on NB p. 9.

Hands-on Science

Coordinating Activities

- **Habitat Diorama** – You will need a shoe box, construction paper, glue, and markers. Have the students choose which habitat they would like to create, for example the arctic, the desert, the grasslands, the forest, or the backyard. Then, have them create that environment inside of a shoe box using construction paper and markers.

Review Sheet

Living World Review Sheet Answers

1. All the options should be circled.
2. Students should have circled: Giraffe, Human Being, Flower
3. Virus
4. Algae, Protozoans, Amoebas
5. Spores
6. Are
7. True

To-Do List

Essential

- ☐ Read the appropriate reading assignment.
- ☐ Define habitat.
- ☐ Complete the lapbook or notebook assignments.
- ☐ Do the "Habitat Diorama" activity.

Optional

- ☐ Get one or more of the library books to read.
- ☐ Fill out a lab report sheet for one of the activities.
- ☐ Complete the Living World Review Sheet. (pp. 118-119)

SCIENCE CHUNKS

Invertebrate Unit

Lesson 1: Mollusks

Information

Reading Assignments

- **Younger Students** – *Kingfisher Encyclopedia of Animals* p. 127 Slug and Snail, p. 131 Octopus
- **Older Students** – *Animal! (DK and Smithsonian)* pp. 24-25 Cephalopods, pp. 26-27 Mollusks

(Optional) Books from the Library

- *Are You a Snail? (Backyard Books)* by Judy Allen
- *Life of the Snail (Nature Watch)* by Theres Buholzer
- *Tiny Snail* by Tammy Carter Bronson
- *An Octopus Is Amazing (Let's-Read-and-Find... Science, Stage 2)* by Patricia Lauber
- *Octopus (Day in the Life: Sea Animals)* by Louise Spilsbury
- *Good Thing You're Not an Octopus!* by Julie Markes and Maggie Smith

Notebooking

Vocabulary

Go over the following words with your students. Then, have them create a flashcard or copy the definitions into the glossary.

- **Invertebrate** – An animal without a backbone. (Flashcard LT p. 27; Glossary NB p. 56)
- **Mollusk** – A group of invertebrate animals, including snails, slugs, and octopuses. (Flashcard LT p. 28; Glossary NB p. 59)

Writing Instructions

- **Lapbook** – Have the students begin the Invertebrate lapbook by cutting out and coloring the cover on LT p. 21. Then, have the students glue the sheet onto the front.
- **Lapbook (multi-week)** – Have the students work on the Mollusk and Crustacean accordion fold-book on LT p. 22. Cut out the page, and color the pictures of the snail and octopus. Have the students tell you what they have learned about the snail and octopus. Then, write their favorite piece of information next to each of the animals on the page.
- **Notebook** – Have the students dictate, copy, or write one to four sentences on what they have learned about mollusks on NB p. 12.

Hands-on Science

Coordinating Activities

✂ **Snail Home** – Have the students attract a mollusk (a snail or slug) to observe. You will need small, unglazed terracotta pot (soaked overnight), and a small rock or piece of mulch. Have the students set a pot that has been soaked in water upside down in the garden. Have them prop up one side with a small rock. Try to set the pot in a shaded place that has a fair amount of foliage nearby. After 4 hours, check the pot to see if any snails or slugs have made their home there. (*If there are no snails, you can try to give it more time by watering the pot and propping it up again. Then, check the pot the next morning to see if there are snails in there this time.*)

✂ **(Optional) Edible Octopus** – Make octopus hot dogs. Begin by cutting a hot dog two-thirds of the way up several times to create legs. Then, cook the hot dog on the stove according to the package instructions. Serve with your student's favorite side!

To-Do List

Essential

- ☐ Read the appropriate reading assignment.
- ☐ Define invertebrate and mollusk.
- ☐ Complete the lapbook or notebook assignments.
- ☐ Do the "Snail Home" activity.

Optional

- ☐ Get one or more of the library books to read.
- ☐ Fill out a lab report sheet (p. 103) for one of the activities.
- ☐ Do the "Edible Octopus" activity

Lesson 2: Crustaceans

Information

Reading Assignments

- **Younger Students** – *Kingfisher Encyclopedia of Animals* p. 131 Shrimp and Prawn, p. 132 Crab
- **Older Students** – *Animal! (DK and Smithsonian)* pp. 28-29 Common Lobster, pp. 30-31 Crustaceans

(Optional) Books from the Library

- *Shrimp (Blastoff! Readers: Oceans Alive)* by Colleen A. Sexton
- *Shrimp (Underwater World)* by Deborah Coldiron
- *A Day on a Shrimp Boat* by Ching Yeung Russell
- *Crab (Welcome Books: Ocean Life)* by Lloyd G. Douglas
- *In One Tidepool: Crabs, Snails, and Salty Tails* by Anthony D. Fredericks
- *Clumsy Crab* by Ruth Galloway

Notebooking

Vocabulary

Go over the following word with your students. Then, have them create a flashcard or copy the definition into the glossary.

- **Shellfish** – An aquatic invertebrate animal having a shell. (Flashcard LT p. 28; Glossary NB p. 63)

Writing Instructions

- **Lapbook (multi-week)** – Have the students complete the Mollusk and Crustacean accordion fold-book on LT p. 22. Have the students color the remaining pictures. Have the students tell you what they have learned about the shrimp and crab. Then, write their favorite piece of information next to each animal on the page. Fold the page, and glue it into the lapbook.
- **Lapbook** – Have the students cut out and color the "Invertebrate" poem on LT p. 23. Once they have finished, have them glue the poem into the lapbook.
- **Notebook** – Have the students dictate, copy, or write one to four sentences on what they have learned about crustaceans on NB p. 13.

Hands-on Science

Coordinating Activities

- **Pincer Pickup** – Have the students pretend to eat like a crustaceans. You will need tweezers, a toothpick, a cotton swab, a gummy worm, a marshmallow, and a lettuce leaf. Lay the gummy worm on a flat surface, and use the cotton swab to try to pick up the worm. Rate how easy (1) to how difficult (10) it is to pick up. Lay the gummy worm on a flat surface, and use the toothpick to try to pick up the worm. Rate how easy (1) to how difficult (10) it is to pick up. Lay the gummy worm on a flat surface, and use the tweezers to try to pick up the worm. Rate how easy (1) to how difficult (10) it is to pick up. Repeat steps 2 to 4 with the lettuce leaf and the marshmallow.
- **(Optional) Microscopic Shrimp** — Examine brine shrimp. Begin by purchasing some brine shrimp from your local pet store. Place them in a shallow dish with a bit of water, and use a magnifying glass to observe them. If you have a microscope, make a quick wet mount slide so that your students can observe the brine shrimp under the microscope. You can make your own wet mount slide by following these directions:
    - https://elementalscience.com/blogs/science-activities/how-to-make-a-microscope-slide

To-Do List

Essential

- ☐ Read the appropriate reading assignment.
- ☐ Define shellfish.
- ☐ Complete the lapbook or notebook assignments.
- ☐ Do the "Pincer Pickup" activity.

Optional

- ☐ Get one or more of the library books to read.
- ☐ Fill out a lab report sheet for one of the activities.
- ☐ Do the "Microscopic Shrimp" activity

Lesson 3: Arachnids and Cnidarians

Information

Reading Assignments

- **Younger Students** – *Kingfisher Encyclopedia of Animals* p. 128 Jellyfish, p. 133 Spider
- **Older Students** – *Animal! (DK and Smithsonian)* pp. 20-21 Cnidarians, pp. 52-53 Arachnids

(Optional) Books from the Library

- *Jellyfish (Ocean Life Up Close: Blastoff Readers, Level 3)* by Christina Leaf
- *Jellyfish! (Penguin Young Readers, Level 4)* by Ginjer L. Clarke
- *Jellyfish (A Day in the Life: Sea Animals)* by Louise Spilsbury
- *Time For Kids: Spiders!* by Nicole Iorio
- *Spinning Spiders (Let's-Read-and-Find... Science 2)* by Melvin Berger
- *The Very Busy Spider* by Eric Carle
- *Diary of a Spider* by Doreen Cronin

Notebooking

Vocabulary

Go over the following word with your students. Then, have them create a flashcard or copy the definition into the glossary.

- **Arthropod** – An animal with a jointed body, such as an insect or spider. (Flashcard LT p. 29; Glossary NB p. 52)

Writing Instructions

- **Lapbook** – Have the students cut out the pages of the Arachnid and Cnidarian mini tab-book on LT p. 24. Ask them to tell you what they have learned about jellyfish and spiders. Write their narrations on each page of the tab-book. Have them color the pictures on each page before assembling the tab-book. Once the students have finished, glue the tab-book into their lapbook.
- **Notebook**– Have the students dictate, copy, or write one to four sentences on what they have learned about arachnids and cnidarians on NB p. 14.

Hands-on Science

Coordinating Activities

- **Web Talk** – Have the students feel what a spider feels when prey lands on its web. You will need about 5 feet of string. Tie one end of the string to something stationary—the top of a chair or a door knob will work. Then, give the other end of the string to the student. Have the student walk far enough away from the object so that the string is taut, and then have the student face away from you. Pluck the string with differing degrees of force, and have the student see if they can feel the differences.

✂ **(Optional) Jellyfish in a Jar** – Have the students make a model jellyfish. You will need 2-Liter bottle with label removed or a glass jar, iridescent film paper, rubber band, scissors, blue food coloring, water, glue, and a skewer. You can find the directions here:

- https://kidsactivitiesblog.com/75076/jellyfish-in-a-bottle/

To-Do List

Essential

- ☐ Read the appropriate reading assignment.
- ☐ Define arthropod.
- ☐ Complete the lapbook or notebook assignments.
- ☐ Do the "Web Talk" activity.

Optional

- ☐ Get one or more of the library books to read.
- ☐ Fill out a lab report sheet for one of the activities.
- ☐ Do the "Jellyfish in a Jar" activity

Lesson 4: Invertebrate Life Cycles

Information

Reading Assignments

- **Younger Students** – *Kingfisher Encyclopedia of Animals* p. 138 Butterfly and Moth, p. 140 Cricket and Grasshopper
- **Younger Students** – "Life Cycles" article p. 106 of this guide
- **Older Students** – *Animal! (DK and Smithsonian)* pp. 40-41 Butterflies and Moths, pp. 252-253 Life Cycles

(Optional) Books from the Library

- *From Caterpillar to Butterfly (Let's-Read-and-Find...)* by Deborah Heiligman
- *National Geographic Readers: Great Migrations Butterflies* by Laura F. Marsh
- *Caterpillars and Butterflies (Usborne Beginners)* by Stephanie Turnbull
- *Life of a Grasshopper (Life Cycles)* by Clare Hibbert
- *Grasshoppers (Bugs Bugs Bugs)* by Margaret Hall
- *Are You a Grasshopper? (Backyard Books)* by Judy Allen

Notebooking

Vocabulary

Go over the following word with your students. Then, have them create a flashcard or copy the definition into the glossary.

- **Metamorphosis** – A process by which an animal's body changes shape during its life cycle. (Flashcard LT p. 29; Glossary NB p. 59)

Writing Instructions

- **Lapbook** – Have the students complete the Metamorphosis Sheet on LT p. 25. Have them color the butterfly life cycle and label the egg, caterpillar, chrysalis, and butterfly. You can use the diagram provided in the lapbooking templates to help your students label the life cycle.

- **Lapbook (multi-week)** – Have the students work on the Insect accordion fold-book on LT p. 26. Cut out the page, and color the pictures of the butterflies and grasshoppers. Have the students tell you what they have learned about the butterflies and grasshoppers. Then, write their favorite piece of information next to each of the animals on the page.

- **Notebook** – Have the students dictate, copy, or write one to four sentences on what they have learned for invertebrate life cycles on NB p. 15.

Hands-on Science

Coordinating Activities

- **Butterfly Life Cycle** – Have the students learn about the butterfly life cycle using the butterfly life cycle picture on p. 107. Have the students cut out and color the butterfly life cycle pictures, or have the kids can draw their own pictures. Have them glue them in order on a separate sheet of paper. Once they are done, have them label the steps with "egg," "caterpillar," "chrysalis," and "butterfly." If you want to see the life cycle in action, you can use a grow a butterfly kit. Here is a post to help with this project:
  - https://elementalscience.com/blogs/science-activities/how-to-grow-a-butterfly-a-zoology-project

- **(Optional) Clothespin Grasshopper** – Have the students make a grasshopper clothespin. Begin by painting a clothespin completely green. While it dries, cut a green pipe cleaner into four sections, and repeat so that you have eight sections. Curl the ends of two of the sections for antennae, and bend the remaining six sections for legs. Once the clothespin is dry, attach the legs and antennae. Finally, add two googly eyes below the antennae to complete your grasshopper.

To-Do List

Essential

- ☐ Read the appropriate reading assignment.
- ☐ Define metamorphosis.
- ☐ Complete the lapbook or notebook assignments.
- ☐ Do the "Butterfly Life Cycle" activity.

Optional

- ☐ Get one or more of the library books to read.
- ☐ Fill out a lab report sheet for one of the activities.
- ☐ Do the "Clothespin Grasshopper" activity

Lesson 5: Insects

Information

Reading Assignments

- **Younger Students** – *Kingfisher Encyclopedia of Animals* p. 134 Ant and Termite, p. 137 Bee and Wasp
- **Older Students** – *Animal! (DK and Smithsonian)* pp. 42-43 Honeybee, pp. 44-45 Insects

(Optional) Books from the Library

- *Ant Cities (Let's-Read-and-Find... Science 2)* by Arthur Dorros
- *National Geographic Readers: Ants* by Melissa Stewart
- *Are You an Ant? (Backyard Books)* by Judy Allen
- *Are You a Bee? (Backyard Books)* by Judy Allen
- *The Life and Times of the Honeybee* by Charles Micucci
- *DK Readers: Busy, Buzzy Bee (Level 1: Beginning to Read)* by Karen Wallace
- *What if There Were No Bees?* by Suzanne Buckingham Slade and Carol Schwartz

Notebooking

Vocabulary

Go over the following word with your students. Then, have them create a flashcard or copy the definition into the glossary.

- **Insect** – An invertebrate animal that has three body parts (head, thorax, and abdomen) and six legs. (Flashcard LT p. 30; Glossary NB p. 55)

Writing Instructions

- **Lapbook (multi-week)** – Have the students complete the Insects accordion fold-book on LT p. 26. Have the students color the remaining pictures. Have the students tell you what they have learned about the ants and bees. Then, write their favorite piece of information next to each animal on the page. Fold the page, and glue it into the lapbook.
- **Notebook** – Have the students dictate, copy, or write one to four sentences on what they have learned for insects on NB p. 16.

Hands-on Science

Coordinating Activities

- **Insect Hunt** – Have the students observe the insects that are in your area. You will need a piece of fruit, honey or syrup, and a plate. Have the students set the piece of fruit on half of the plate and squirt a bit of the honey or syrup on the other half. Next, set the plate on the ground outside in a place that the students can easily see from the inside. 4. Let them observe the plate over the next hour or so. (*You can have the students record the insects they see on the box on a separate sheet of paper.*)

✂ **(Optional) Ant Farm** – Have the students build your own ant farm using the directions from the following website:

🖰 http://tlc.howstuffworks.com/family/ant-activities1.htm

Review Sheet

Invertebrates Review Sheet Answers

1. Without
2. Both in land and in water
3. 8
4. Pincers
5. True
6. 8
7. 3, 6
8. True
9. Life cycle
10. False (*Caterpillars turn into butterflies through metamorphosis.*)
11. Hard

To-Do List

Essential

☐ Read the appropriate reading assignment.

☐ Define insect.

☐ Complete the lapbook or notebook assignments.

☐ Do the "Insect Hunt" activity.

Optional

☐ Get one or more of the library books to read.

☐ Fill out a lab report sheet for one of the activities.

☐ Do the "Ant Farm" activity

☐ Complete the Invertebrates Review Sheet. (pp. 120-121)

SCIENCE CHUNKS

Planet Earth Unit

Lesson 1: Inside the Earth

Information

Reading Assignments

- **Younger Students** – *DK Children's Encyclopedia* p. 135 Inside Earth
- **Older Students** – *Kingfisher Science Encyclopedia* pp. 8-9 Earth's Structure

(Optional) Books from the Library

- *Flip The Flaps: Planet Earth* by Mike Goldsmith and Nicki Palin
- *The Magic School Bus Inside the Earth (Magic School Bus)* by Joanna Cole
- *See Inside Planet Earth (Usborne Flap Book)* by Katie Daynes and Peter Allen

Notebooking

Vocabulary

Go over the following word with your students. Then, have them create a flashcard or copy the definition into the glossary.

- **Magma** – Molten rock found under earth's surface. (Flashcard LT p. 40; Glossary NB p. 57)

Writing Instructions

- **Lapbook** – Have the students begin the Plant Earth lapbook by cutting out and coloring the cover on LT p. 33. Then, have the students glue the sheet onto the front.
- **Lapbook** – Have the students begin the Labeled Earth Sheet on LT p. 34 by adding labels for "core," "mantle," and "crust" on the inner portion of the Earth picture. (*See the completed sheet on LT p. 34 for the answers.*) Then, have them set the sheet aside for next week.
- **Notebook** – Have the students dictate, copy, or write three to five sentences about the inside of the Earth on NB p. 18. (*You can also have the students label the different parts of the Earth on the picture.*)

Hands-on Science

Coordinating Activities

- **Model Earth** – You will need modeling (or air dry) clay (red, orange, yellow, blue, and green) and a ruler. Have the students begin by making a ball about ½ inch in diameter out of the yellow clay. (*This represents the Earth's inner core.*) Then, have the students make a flat circle out of the red clay and wrap it around the ball so that the ball is about 1 inch across. (*This layer represents the Earth's outer core.*) Next, have the students make a flat circle out of the orange clay and wrap it around the ball so that the ball is about 2 inches across. (*This layer represents the Earth's mantle, the thickest layer.*) Finally, have the students make a flat circle out of the blue clay and wrap it around the ball so that the ball is about 2 ½ inches across. Then, use the green clay to make a few flattened pieces to represent the continents, and layer them over your ball. (*The blue and green clay layers represent the Earth's crust.*) After the students have finished the ball, cut the ball in half and have them observe the layers.

(Optional) Edible Earth – Make an edible earth core with the students using Rice Krispie treats and icing. See the following website for directions:

http://teachbesideme.com/geography-earths-core-project/

To-Do List

Essential

- ☐ Read the appropriate reading assignment.
- ☐ Define magma.
- ☐ Complete the lapbook or notebook assignments.
- ☐ Do the "Model Earth" activity.

Optional

- ☐ Get one or more of the library books to read.
- ☐ Fill out a lab report sheet (p. 103) for one of the activities.
- ☐ Do the "Edible Earth" activity.

Lesson 2: On the Earth

Information

Reading Assignments

- **Younger Students** – *DK Children's Encyclopedia* p. 77 Day and Night, p. 83 Earth
- **Older Students** – *Kingfisher Science Encyclopedia* p. 2 Earth and the Solar System, p. 3 Earth's Rotation

(Optional) Books from the Library

- *Day and Night (First Step Nonfiction: Discovering Nature's Cycles)* by Robin Nelson
- *Day and Night (Patterns in Nature)* by Margaret Hall and Jo Miller
- *What Makes Day and Night (Let's-Read-and-Find... Science 2)* by Franklyn M. Branley
- *The Seven Continents (Rookie Read-About Geography)* by Will Mara
- *Continents in My World* by Ella Cane and Gail Saunders-Smith
- *Planet Earth/Inside Out* by Gail Gibbons

Notebooking

Vocabulary

Go over the following word with your students. Then, have them create a flashcard or copy the definition into the glossary.

- **Continent** – A large area of land on Earth. (Flashcard LT p. 41; Glossary NB p. 53)

Writing Instructions

- **Lapbook** – Have the students finish the Labeled Earth Sheet by adding labels for "ocean" and "continent" on the outer portion of the Earth picture. (*See the completed sheet on LT p. 34 for the answers.*) Then, have them glue the sheet into the lapbook.
- **Lapbook** – Have the students cut out and color the "Seven Continents" poem on LT p. 35. Once they have finished, have them glue the poem into the lapbook.
- **Notebook** – Have the students dictate, copy, or write two to six sentences on what they have learned about the Earth on NB p. 14.

Hands-on Science

Coordinating Activities

- **Day or Night** – Have the students see the day/night cycle in action. You will need a globe (or large ball), a desk lamp, and a Post-it tab (or another type of removable marker). Have the students mark where they live on the globe with the Post-it tab. Then, have them place the globe on a desk and shine the desk lamp on the portion of the globe where they live. Have the students slowly spin (or rotate) the globe to see what happens. (*The students should see that as they rotate the globe, the place where they live rotates out of the light. As they continue to spin the globe, the place where they live will rotate back into the light – creating a full day/night cycle.*)

(Optional) Continents – Have the students label and color the continents (North America, South American, Europe, Asia, Africa, Australia, and Antarctica) on a map of the Earth. You can use the map template provided in the appendix p. _ and the answers are provided here for your reference. After you do that, memorize the continents with the help of the following song:

- "The Continent Song" - https://www.youtube.com/watch?v=fqsCWZtZlk4

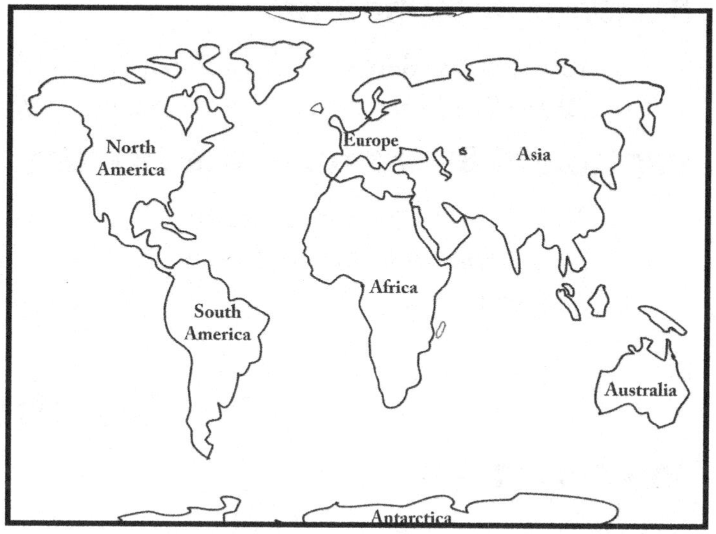

To-Do List

Essential

- ☐ Read the appropriate reading assignment.
- ☐ Define continent.
- ☐ Complete the lapbook or notebook assignments.
- ☐ Do the "Day or Night" activity.

Optional

- ☐ Get one or more of the library books to read.
- ☐ Fill out a lab report sheet for one of the activities.
- ☐ Do the "Continents" activity.

Lesson 3: Mountains

Information

Reading Assignments

- **Younger Students** – *DK Children's Encyclopedia* p. 172 Mountains
- **Older Students** – *Kingfisher Science Encyclopedia* pp. 22-23 Building Mountains

(Optional) Books from the Library

- *Mountains* by Seymour Simon
- *How Mountains Are Made (Let's-Read-and-Find Out Science 2)* by Kathleen Zoehfeld and James Hale

Notebooking

Vocabulary

Go over the following word with your students. Then, have them create a flashcard or copy the definition into the glossary.

- **Mountain Range** – A row of mountains that form a sort of chain across the land. (Flashcard LT p. 41; Glossary NB p. 60)

Writing Instructions

- **Lapbook** – Have the students complete the Mountains shutterfold book on LT p. 36. Have them cut out and color the page. Next, have the students tell you what they have learned about mountains and write it on the inside of the booklet. Then, have them fold the book and glue it into the lapbook.
- **Notebook** – Have the students dictate, copy, or write two to six sentences on what they have learned about mountains on NB p. 20.

Hands-on Science

Coordinating Activities

- **Towel Mountain** – Have the students see how mountains form by using a large bath towel and a flat, smooth surface. Start by folding the towel in half and then fold it in half again, so that you have a thick, but long rectangle. Lay the towel on a flat, smooth surface, such as a counter top or table. Have the students place their hands on both of the short ends of the rectangle. Then, have them slowly try to bring their hands together so that the towel buckles, forming peaks along the rectangle. (*Mountains form in much the same way. Heat and pressure soften the rock and when they plates meet, this can cause the plates to buckle or forces one plate to bend under the other, creating mountain ranges.*)

✂ **(Optional) Mountain Model** – Have the students make their own model of a mountain. You will need cardboard, air-dry clay or salt dough, and paint (white, gray, green, and blue). Have the students begin by making a mountain out of air-dry clay or salt dough on a cardboard sheet. Set their creations aside to dry. Once the mountains are dry, discuss the parts of a mountain with them using the following information:

1. Peak - The very top point of the mountain.
2. Alpine Zone - The coldest region of the mountain, where snow can be found year-round depending upon where the mountain is located on the globe.
3. Snowline - The line at which the snow ends.
4. Subalpine Zone - The region between the snowline and the treeline. A few cold hard plants grow in this region.
5. Treeline - The line at which the trees begin to grow.

Then, have the students paint and label these ares on their mountains. They can paint the peak and the alpine zone white, the subalpine zone grey, and everything below the treeline green. Then, the students can add ribbons of blue from the snowline down to represent rivers and streams.

To-Do List

Essential

- ☐ Read the appropriate reading assignment.
- ☐ Define mountain range.
- ☐ Complete the lapbook or notebook assignments.
- ☐ Do the "Towel Mountain" activity.

Optional

- ☐ Get one or more of the library books to read.
- ☐ Fill out a lab report sheet for one of the activities.
- ☐ Do the "Mountain Model" activity.

Lesson 4: Maps

Information

Reading Assignments

- **Younger Students** – *DK Children's Encyclopedia* p. 72 Compass, p. 155 Maps
- **Older Students** – *Kingfisher Science Encyclopedia* pp. 46-47 Maps and Mapping

(Optional) Books from the Library

- *There's a Map on My Lap!: All About Maps* by Tish Rabe and Aristides Ruiz
- *Pete the Cat and the Treasure Map* by James Dean
- *The Equator (Where on Earth? Mapping Parts of the World)* by Todd Bluthenthal
- *Me On The Map (Rise and Shine)* by Joan Sweeney
- *Maps and Globes* by Jack Knowlton and Harriet Barton

Notebooking

Vocabulary

Go over the following word with your students. Then, have them create a flashcard or copy the definition into the glossary.

- **Equator** – The most important line of latitude, which divides the Earth in half. (Flashcard LT p. 42; Glossary NB p. 54)

Writing Instructions

- **Lapbook** – Have the students cut out the pages and color the cover of the Maps Mini Tab-book on LT p. 37. Then, have the students color on the map. Have them tell you what they have learned about maps and write it for them on the on the "Maps" page. After that, have the students draw a map of their room or neighborhood on the "My Map" page. Lastly, staple the pages together and glue the biome flip-book into the lapbook.
- **Lapbook** – Have the students add the Compass memory sentence to the lapbook. Have them cut out and color the poem sheet found on LT p. 38. Once they are done, have them glue the sheet into the lapbook.
- **Notebook** – Have the students dictate, copy, or write two to six sentences on what they have learned about maps and draw a map of their room or neighborhood on NB p. 21.

Hands-on Science

Coordinating Activities

- **Homemade Compass** – Have the students make their own compass. You will need a straight pin, a milk jug, a knife, and a magnet. You can find the directions for this here:

https://elementalscience.com/blogs/science-activities/how-to-make-a-compass

(Optional) Where in the World Game – Have the students practice finding locations on the globe. You will need access to a globe or a flat map. (*You can get a blank one in the appendix on p. 108.*) Then call out a location, and have the students point to it on the globe or map with your help. Start with where you are located, then you can all out the following:

- Places you have visited
- Places you used to live
- Places where your family lives
- Places you hope to visit one day
- Places in the news
- Places you have read about
- Places where the students favorite animals are located
- And so on

Be sure that you call out at least one place on each continent.

To-Do List

Essential

- ☐ Read the appropriate reading assignment.
- ☐ Define equator.
- ☐ Complete the lapbook or notebook assignments.
- ☐ Do the "Homemade Compass" activity.

Optional

- ☐ Get one or more of the library books to read.
- ☐ Fill out a lab report sheet for one of the activities.
- ☐ Play the "Where in the World" game.

Lesson 5: Conservation

Information

Reading Assignments

- **Younger Students** – *DK Children's Encyclopedia* p. 72 Conservation, p. 205 Recycling
- **Older Students** – *Kingfisher Science Encyclopedia* pp. 462-463 Conservation in Action

(Optional) Books from the Library

- *Where Does the Garbage Go?: Revised Edition (Let's-Read-and-Find Out About Science)* by Paul Showers and Randy Chewning
- *The Three R's: Reuse, Reduce, Recycle (What Do You Know About? Books)* by Nuria Roca and Rosa M. Curto

Notebooking

Vocabulary

Go over the following word with your students. Then, have them create a flashcard or copy the definition into the glossary.

- **Recycling** – The act of turning waste, like used cans, bottles, and paper, into new, usable materials. (Flashcard LT p. 42; Glossary NB p. 62)

Writing Instructions

- **Lapbook** – Have the students complete the Caring for Earth mini-book on LT p. 39. Have them cut out the template and color it. Next, have the students tell you what they have learned about conservation and recycling. Then, have them fold the book in half and glue the booklet into the lapbook.
- **Notebook** – Have the students dictate, copy, or write two to six sentences on what they have learned about conservation on NB p. 22.

Hands-on Science

Coordinating Activities

- **Recycling Plan** – Have the students practice recycling at home by creating a recycling plan for your family. You can visit the following website for more information on recycling and ideas on how to set up a recycling plan.
  - http://elementalblogging.com/homeschool-science-corner-recycling-plastics/
  - http://www.bravenewleaf.com/environment/2007/11/how-to-set-up-a.html

Review Sheet

Planet Earth Review Sheet Answers

1. C, B, A
2. The rotation of the Earth
3. True
4. Range
5. Can
6. True
7. False (*There are many different kinds of maps.*)
8. All the answers should be circled

To-Do List

Essential

- ☐ Read the appropriate reading assignment.
- ☐ Define recycling.
- ☐ Complete the lapbook or notebook assignments.
- ☐ Do the "Recycling Plan" activity.

Optional

- ☐ Get one or more of the library books to read.
- ☐ Fill out a lab report sheet for one of the activities.
- ☐ Complete the Planet Earth Review Sheet. (pp. 122-123)

SCIENCE CHUNKS

Periodic Table Unit

Lesson 1: The Elements and The Periodic Table

Information

Reading Assignments

- **Younger Students** – *Basher Science Periodic Table* p. 6 Periodic Table
- **Older Students** – *DK Eyewitness The Elements* pp. 8-9 The Periodic Table, pp. 10-11 Element Groups and Sets

  Note—On the right-hand side the periodic table, things begin to divert a bit. Some resources will group the elements according to the element at the top of the column, that is, Boron Group, Oxygen Group, and so on. Other resources will group the elements according to similar properties, that is, Metalloids, Nonmetals, and so on. Although both methods are technically correct, in this unit, we have chosen to go with the first method of grouping the elements based on similar characteristics. Here is an article about why these differences occur:

  - https://elementalscience.com/blogs/news/the-periodic-table

(Optional) Books from the Library

- *The Elements (True Books)* by Matt Mullins
- *Elements and Compounds (Building Blocks of Matter)* by Louise Spilsbury and Richard Spilsbury
- *The Periodic Table (True Books: Elements)* by Salvatore Tocci

Notebooking

Vocabulary

Go over the following words with your students. Then, have them create a flashcards or copy the definitions into the glossary.

- **Element** – A substance made up of one type of atom, which cannot be broken down by chemical reaction to form a simpler substance. (Flashcard LT p. 61; Glossary NB p. 54)
- **Periodic Table** – A systematic arrangement of the elements in order of increasing atomic number. (Flashcard LT p. 62; Glossary NB p. 61)

Writing Instructions

- **Lapbook** – Have the students begin the Periodic Table lapbook by cutting out and coloring the cover on LT p. 45. Then, have the students glue the sheet onto the front.
- **Lapbook** – Have the students complete the Elements Diagram Sheet on LT p. 46. Have the students cut it out and color it. Then, have them label the atomic number, symbol, and

atomic mass on the element. Finally, have them glue the sheet into the lapbook.

- **Lapbook** – Have the students work on the Periodic Table Book on LT p. 47. Have the students cut out, fold the template, and color the picture on the cover. Have the students write their narration about the periodic table inside the mini-book. Then glue the mini-book into the lapbook.
- **Notebook** – Have the students dictate, copy, or write one to four sentences on what they have learned about the elements and the periodic table on NB p. 24.

Hands-on Science

Coordinating Activities

- Table Sorting – Have the students deepen their understanding of the periodic table by doing a table sorting activity. You will need LEGOs in a variety of colors and sizes (You can also used stuffed animals, buttons, beads, or any other object with different sizes and colors if you don't have any LEGOs), paper, and pen. Place the LEGOs in an unorganized pile in front of the students and help them draw a 4 by 6 grid on the piece of paper. (If you are using larger objects to sort, such as stuffed animals, you can create this grid on the floor with masking tape.) Explain to the students that you are going to make a periodic table of LEGOs. In your table, the LEGOs are going to get bigger as you go down the grid and darker as you go across. (See the included grid below for visual explanation.) Have the students sort the LEGOs by size and color onto the grid. As they sort, share with them how the periodic table in chemistry is an organized assortment of elements set up in a grid, similar to how they are sorting their LEGOs.
- **(Optional) Periodic Table Game** – Have the students play a game of Periodic Table Battleship! You can see directions and print out game sheets at the following website:
  - http://teachbesideme.com/periodic-table-battleship/

	White	Yellow	Red	Blue	Brown	Black
smallest						
largest						

To-Do List

Essential

- ☐ Read the appropriate reading assignment.
- ☐ Define element and periodic table.
- ☐ Complete the lapbook or notebook assignments.
- ☐ Do the "Table Sorting" activity.

Optional

- ☐ Get one or more of the library books to read.
- ☐ Fill out a lab report sheet (p. 103) for one of the activities.
- ☐ Do the "Periodic Table Game" activity

Lesson 2: Alkali Metals

Information

Reading Assignments

- **Younger Students** – *Basher Science Periodic Table* p. 10 Alkali Metals, p. 8 Hydrogen, p. 14 Sodium
- **Older Students** – *DK Eyewitness The Elements* pp. 14-17 Alkali Metals

(Optional) Books from the Library

- *The Alkali Metals: Lithium, Sodium, Potassium, Rubidium, Cesium, Francium (Understanding the Elements of the Periodic Table)* by Kristi Lew
- *Hydrogen and the Noble Gases (True Books: Elements)* by Salvatore Tocci
- *Hydrogen: Running on Water (Energy Revolution)* by Niki Walker
- *Sodium (Elements)* by Anne O'Daly
- *Sodium (True Books: Elements)* by Salvatore Tocci

Notebooking

Vocabulary

Go over the following word with your students. Then, have them create flashcard or copy the definition into the glossary.

- **Reactive** – The tendency of a substance to react with other substances. (Flashcard LT p. 62; Glossary NB p. 62)

Mini-book Assembly Instructions

- **Lapbook** – Have the students work on the Central Periodic Table on LT p. 49. Have the students cut out the periodic table template and glue it into their lapbook. Then, color the alkali metals red.
- **Lapbook** – Have the students work on the Alkali Flip-book on LT p. 48. Have the students cut along the solid lines and color the group on the cover red. Next, have the students write several characteristics of alkali metals on the characteristics tab, as well as several facts about hydrogen and sodium on the respective tabs. Then, have them line the pages up and staple the sheets together. Finally, have them glue the flip-book into the lapbook.
- **Notebook** – Have the students dictate, copy, or write one to four sentences about alkali metals, hydrogen, and sodium on NB p. 25.

Hands-on Science

Coordinating Activity

- **Sodium Reaction** – Have the students learn about the chemistry of an alkali metal compound in your kitchen – baking soda, a.k.a. sodium bicarbonate. Add 1 Tablespoon of baking soda to a cup. Then, have the students add a few drops of white vinegar and observe what happens! (This is the classic acid (vinegar) and base (baking soda) reaction. The bubbles you see are a release of the energy and the products from the reaction—carbon dioxide gas, sodium acetate, and water.)
- **(Optional) Which one freezes first?** – Have the students learn about the freezing point of water and how it can be affected by an alkali metal. You will need 3 cups, water, food coloring, and salt. Directions for this project can be found here:
    - https://elementalscience.com/blogs/science-activities/83744963-which-one-freezes-first

To-Do List

Essential

- ☐ Read the appropriate reading assignment.
- ☐ Define reactive.
- ☐ Complete the lapbook or notebook assignments.
- ☐ Do the "Sodium Reaction" activity.

Optional

- ☐ Get one or more of the library books to read.
- ☐ Fill out a lab report sheet for one of the activities.
- ☐ Do the "Which one freezes first?" activity

Lesson 3: Alkaline Earth Metals

Information

Reading Assignments

- 🕮 **Younger Students** – *Basher Science Periodic Table* p. 22 Alkaline Earth Metals, p. 26 Magnesium, p. 28 Calcium
- 🕮 **Older Students** – *DK Eyewitness The Elements* pp. 18-21 (Alkaline Earth Metals)

(Optional) Books from the Library

- 🕮 *The Alkaline Earth Metals: Beryllium, Magnesium, Calcium, Strontium, Barium, Radium (Understanding the Elements of the Periodic Table)* by Bridget Heos
- 🕮 *Calcium (True Books: Elements)* by Salvatore Tocci
- 🕮 *Magnesium (The Elements)* by Colin Uttley

Notebooking

Vocabulary

Go over the following word with your students. Then, have them create flashcard or copy the definition into the glossary.

- **Solution** – A mixture that consists of a substance dissolved in a liquid. (Flashcard LT p. 63; Glossary NB p. 63)

Mini-book Assembly Instructions

- **Lapbook** – Have the students work on the Central Periodic Table. This week, have the students color the alkaline earth metals orange.
- **Lapbook** – Have the students work on the Alkaline Earth Flip-book on LT p. 50. Have the students cut along the solid lines and color the group on the cover orange. Next, have the students write several characteristics of alkaline earth metals on the characteristics tab as well as several facts about magnesium and calcium on the respective tabs. Then, line the pages up and staple the sheets together. Finally, glue the flip-book into the lapbook.
- **Notebook** – Have the students dictate, copy, or write one to four sentences about alkaline earth metals, magnesium, and calcium on NB p. 26.

Hands-on Science

Coordinating Activity

- ✂ **Magnesium Solutions** – Have the students make a magnesium solution and watch what happens. You will need Epsom salts, ammonia, water, and a clear cup. Have the students fill the cup about halfway with warm water. Then, add a teaspoon of Epsom salts and stir until it's dissolved. Next, pour in 2 teaspoons of ammonia, but DO NOT stir. Let the jar sit for 5 minutes, and observe what happens. (The students should see a milky, white substance within 5 minutes. This substance is magnesium hydroxide, which does not dissolve in water. This is a compound of an alkaline earth metal, magnesium.) **Note**—If the cup sits undisturbed for about 30 minutes longer, the white substance will settle out on top of the water layer.

✂ **(Optional) Disappearing Calcium** – Have the students dissolve some calcium! You will need an egg, white vinegar, and a clear cup. Place the egg in the glass and cover it with vinegar. Cover the glass, set it aside, and wait for twenty-four hours. The next day, pour out the vinegar and observe the changes to the shell. (The shell of the egg contains calcium, which is dissolved by the acid in vinegar.)

To-Do List

Essential

- ☐ Read the appropriate reading assignment.
- ☐ Define solution.
- ☐ Complete the lapbook or notebook assignments.
- ☐ Do the "Magnesium Solutions" activity.

Optional

- ☐ Get one or more of the library books to read.
- ☐ Fill out a lab report sheet for one of the activities.
- ☐ Do the "Disappearing Calcium" activity

Lesson 4: Transition Metals

Information

Reading Assignments

- **Younger Students** – *Basher Science Periodic Table* p. 34 Transition Elements, p. 44 Iron, p. 48 Copper
- **Older Students** – *DK Eyewitness The Elements* p. 22 Transition Metal, p. 24 Iron, p. 26 Copper

(Optional) Books from the Library

- *The Transition Elements: The 37 Transition Metals (Understanding the Elements of the Periodic Table)* by Mary-Lane Kamberg
- *Iron (Elements)* by Giles Sparrow
- *From Iron to Car (Start to Finish, Second Series)* by Shannon Zemlicka
- *The Story of Iron (First Book)* by Karen Fitzgerald
- *Copper (The Elements)* by Richard Beatty

Notebooking

Vocabulary

Go over the following words with your students. Then, have them create flashcards or copy the definitions into the glossary.

- **Alloy** – A mixture of two or more metals or a metal and a non-metal. (Flashcard LT p. 63; Glossary NB p. 52)
- **Metal** – The largest class of elements, usually they are shiny and solid at room temperature. (Flashcard LT p. 64; Glossary NB p. 58)

Mini-book Assembly Instructions

- **Lapbook** – Have the students work on the Central Periodic Table. This week, have the students color the transition metals yellow.
- **Lapbook** – Have the students work on the Transition Metals Tab-book on LT pp. 51-52. Have the students cut along the solid lines and color the group on the cover yellow. Next, have the students write several characteristics of transition metals on the characteristics tab, as well as several facts about iron and copper on the respective tabs. Then, have them line the pages up and staple the sheets together. Finally, glue the tab-book into the lapbook.
- **Notebook** – Have the students dictate, copy, or write one to four sentences about transition metals, iron, and copper on NB p. 27.

Hands-on Science

Coordinating Activity

- **Rusted** – Have the students see what happens to a transition metal when it is exposed to the air. You will need steel wool, vinegar, and a jar with lid. Have the students place the steel wool in one of the jars and cover it with vinegar. Put the lid on the jar, and let the steel wool soak for 2 hours. (This is done to remove any coating that may be on the steel wool so that the iron could react with the moist air.) After 2 hours, have the students take the steel wool out, and set it on a the jar lid. Let it sit undisturbed for 30 minutes. After a half an hour has passed, observe the changes in the steel wool. (The students should see that the steel wool has turned a brownish-red color. This is because iron rusts, or oxidizes, when it is exposed to moist air. This is why iron we use in daily life is often coated with other metals or mixed with other elements to form rust-resistant alloys.)
- **(Optional) Metal Plating** – Have the students turn an iron nail into one covered with copper. You will need white vinegar, salt, six pennies, glass cup, and two iron nails. Have the students cover the bottom of the cup with a thin layer of salt. Then, add the pennies, and cover them with the vinegar. After 10 minutes, take the pennies out, and set them on a paper towel. Be sure to reserve the vinegar mixture. Add one of the nails to the vinegar solution, and leave the other one next to the glass. Let both of them sit undisturbed for 45 minutes. Take the nail out of the vinegar solution, and compare it to the nail that was left outside the glass. (The students should see that the iron nail from the vinegar solution has a thin, shiny, brownish coating on it. The coating may or may not be completely covering the nail. After the pennies were removed, the vinegar solution contained copper ions. Over time, they adhered to the iron nail, giving it a thin, shiny, brownish coating of copper. This is a small glimpse of how metal (copper) plating works.)

To-Do List

Essential

- ☐ Read the appropriate reading assignment.
- ☐ Define alloy and metal.
- ☐ Complete the lapbook or notebook assignments.
- ☐ Do the "Rusted" activity.

Optional

- ☐ Get one or more of the library books to read.
- ☐ Fill out a lab report sheet for one of the activities.
- ☐ Do the "Metal Plating" activity

Lesson 5: Boron Elements

Information

Reading Assignments

- **Younger Students** – *Basher Science Periodic Table* p. 76 Boron Elements, p. 78 Boron, p. 80 Aluminum
- **Older Students** – *DK Eyewitness The Elements* pp. 46-49 The Boron Group

(Optional) Books from the Library

- *The Boron Elements: Boron, Aluminum, Gallium, Indium, Thallium (Understanding the Elements of the Periodic Table)* by Heather Hasan
- *Aluminum* by Heather Hasan
- *Boron (Elements)* by Richard Beatty

Notebooking

Vocabulary

Go over the following word with your students. Then, have them create flashcard or copy the definition into the glossary.

- **Metalloid** – An element that shares some of the properties of metals and nonmetals. (Flashcard LT p. 64; Glossary NB p. 59)

Mini-book Assembly Instructions

- **Lapbook** – Have the students work on the Central Periodic Table. This week, have the students color the boron elements light green.
- **Lapbook** – Have the students work on the Boron Elements Flip-book on LT p. 53. Have the students cut along the solid lines and color the group on the cover light green. Next, have the students write several characteristics of boron elements on the characteristics tab as well as several facts about aluminum and boron on the respective tabs. Then, have them line the pages up and staple the sheets together. Finally, glue the flip-book into the lapbook.
- **Notebook** – Have the students dictate, copy, or write one to four sentences about boron elements, aluminum, and boron on NB p. 28.

Hands-on Science

Coordinating Activity

- **Boron Slime** – Have the students make slime using borax, a boron-containing compound.

You will need gel glue, water, and Borax, which can be found in the laundry aisle. In a plastic baggie, have the students mix equal parts of glue and water. Meanwhile, in a cup, mix a quarter cup of water with half a teaspoon of Borax. Then, add the Borax solution to the baggie and have the students massage the bag for a few minutes until a nice firm slime has formed.

✂ **(Optional) Aluminum Video** – Have the students watch a neat reaction where aluminum foil gets dissolved. The video can be viewed at the following link:

- https://www.youtube.com/watch?v=AAKdc7PO3J0

To-Do List

Essential

- ☐ Read the appropriate reading assignment.
- ☐ Define metalloid.
- ☐ Complete the lapbook or notebook assignments.
- ☐ Do the "Boron Slime" activity.

Optional

- ☐ Get one or more of the library books to read.
- ☐ Fill out a lab report sheet for one of the activities.
- ☐ Watch the aluminum video.

Lesson 6: Carbon Elements

Information

Reading Assignments

- **Younger Students** – *Basher Science Periodic Table* p. 86 Carbon Elements, p. 88 Carbon, p. 93 Tin
- **Older Students** – *DK Eyewitness The Elements* pp. 50-53 The Carbon Group

(Optional) Books from the Library

- *The Carbon Elements: Carbon, Silicon, Germanium, Tin, Lead (Understanding the Elements of the Periodic Table)* by Brian Belval
- *Carbon* by Linda Saucerman
- *Carbon (True Books: Elements)* by Salvatore Tocci
- *Tin (True Books: Elements)* by Salvatore Tocci
- *The Invention of the Silicon Chip: A Revolution in Daily Life* by Windsor Chorlton

Notebooking

Vocabulary

Go over the following word with your students. Then, have them create flashcard or copy the definition into the glossary.

- **Nonmetal** – A class of elements that can be non-shiny solids or gases. (Flashcard LT p. 65; Glossary NB p. 60)

Mini-book Assembly Instructions

- **Lapbook** – Have the students work on the Central Periodic Table. This week, have the students color the carbon elements dark green.
- **Lapbook** – Have the students work on the Carbon Elements Flip-book on LT p. 54. Have the students cut along the solid lines and color the group on the cover dark green. Next, have the students write several characteristics of carbon elements on the characteristics tab as well as several facts about carbon and tin on the respective tabs. Then, line the pages up and staple the sheets together. Finally, glue the flip-book into the lapbook.
- **Notebook** – Have the students dictate, copy, or write one to four sentences about carbon elements, carbon, and tin on NB p. 29.

Hands-on Science

Coordinating Activity

- **Carbon Test** – Have the students test for the presence of carbon in rocks. Have them choose a rock, such as limestone, place it in a cup, and cover the rock with white vinegar. Have the students observe what happens. (If there is the presence of bubbles coming from the rock, the carbon-containing compound calcium carbonate present.)
- **(Optional) Tin Can Lights** – Have the students make a tin can luminary. You will need a tin can, a nail, permanent marker, hammer, and a candle. You can find the directions for this project from the following website:
  - http://www.thechaosandtheclutter.com/archives/tin-can-luminaries

To-Do List

Essential

- ☐ Read the appropriate reading assignment.
- ☐ Define nonmetal.
- ☐ Complete the lapbook or notebook assignments.
- ☐ Do the "Carbon Test" activity.

Optional

- ☐ Get one or more of the library books to read.
- ☐ Fill out a lab report sheet for one of the activities.
- ☐ Do the "Tin Can Light" activity.

Lesson 7: Nitrogen Elements

Information

Reading Assignments

- **Younger Students** – *Basher Science Periodic Table* p. 96 Nitrogen Elements, p. 98 Nitrogen, p. 100 Phosphorus
- **Older Students** – *DK Eyewitness The Elements* pp. 54-57 The Nitrogen Group

(Optional) Books from the Library

- *The Nitrogen Elements (Understanding the Elements of the Periodic Table)* by Greg Roza
- *Nitrogen (True Books: Elements)* by Salvatore Tocci
- *Nitrogen* by Heather Hasan
- *Phosphorus (Elements)* by Richard Beatty

Notebooking

Vocabulary

Go over the following word with your students. Then, have them create flashcard or copy the definition into the glossary.

- **Essential Element** – An element that is essential to life on earth, such as carbon, hydrogen, nitrogen, and oxygen. (Flashcard LT p. 65; Glossary NB p. 54)

Mini-book Assembly Instructions

- **Lapbook** – Have the students work on the Central Periodic Table. This week, have the students color the nitrogen elements light blue.
- **Lapbook** – Have the students Have the students work on the Nitrogen Elements Flip-book on LT p. 55. Have the students cut along the solid lines and color the group on the cover light blue. Next, have the students write several characteristics of nitrogen elements on the characteristics tab, as well as several facts about nitrogen and phosphorus on the respective tabs. Then, line the pages up and staple the sheets together. Finally, glue the flip-book into the lapbook.
- **Notebook** – Have the students dictate, copy, or write one to four sentences about nitrogen elements, nitrogen, and phosphorus on NB p. 30.

Hands-on Science

Coordinating Activity

- **Dirty Pennies** – Have the students clean a few pennies with a nitrogen element--phosphorus.

You will need a can of dark cola soda, glass, and several dirty pennies. Have the students place several dirty pennies in a cup. Then, pour enough cola into the cup to cover the pennies. Set the cup aside and let it sit undisturbed overnight. The next morning, fish out the pennies and have the students observe the differences. When you are done, pour the cola down the drain. **Note**—Do not drink the cola as it now has copper ions in it! (Dark cola has phosphoric acid in it, something that gives the soda its tangy flavor. This acid also breaks up the copper-oxygen compound that, with time, makes pennies dark and dull.)

✂ **(Optional) Cycles** – Have the students learn more about the nitrogen cycle by watching the following video:

https://www.youtube.com/watch?v=ZaFVfHftzpI

If you would like for your students to also learn about the phosphorus cycle, you can have them watch the following video:

https://www.youtube.com/watch?v=wdAzQSuypCk

To-Do List

Essential

- ☐ Read the appropriate reading assignment.
- ☐ Define essential element.
- ☐ Complete the lapbook or notebook assignments.
- ☐ Do the "Dirty Pennies" activity.

Optional

- ☐ Get one or more of the library books to read.
- ☐ Fill out a lab report sheet for one of the activities.
- ☐ Do the "Cycles" activity.

Lesson 8: Oxygen Elements

Information

Reading Assignments

- 🕮 **Younger Students** – *Basher Science Periodic Table* p. 106 Oxygen Elements, p. 108 Oxygen, p. 110 Sulfur
- 🕮 **Older Students** – *DK Eyewitness The Elements* pp. 59-61 The Oxygen Group

(Optional) Books from the Library

- 🕮 *The Oxygen Elements: Oxygen, Sulfur, Selenium, Tellurium, Polonium (Understanding the Elements of the Periodic Table)* by Laura La Bella
- 🕮 *Nonmetals (Material Matters/Freestyle Express)* by Carol Baldwin
- 🕮 *Oxygen (True Books: Elements)* by Salvatore Tocci
- 🕮 *Sulfur (The Elements)* by Richard Beatty

Notebooking

Vocabulary

Go over the following word with your students. Then, have them create flashcard or copy the definition into the glossary.

- **Oxidation** – A chemical reaction in which a substance combines with oxygen. (Flashcard LT p. 66; Glossary NB p. 60)

Mini-book Assembly Instructions

- **Lapbook** – Have the students work on the Central Periodic Table. This week, have the students color the oxygen elements dark blue.
- **Lapbook** – Have the students work on the Oxygen Elements Flip-book on LT p. 56. Have the students cut along the solid lines and color the group on the cover dark blue. Next, have the students write several characteristics of oxygen elements on the characteristics tab, as well as several facts about oxygen and sulfur on the respective tabs. Then, line the pages up and staple the sheets together. Finally, glue the flip-book into the lapbook.
- **Notebook** – Have the students dictate, copy, or write one to four sentences about oxygen elements, oxygen, and sulfur on NB p. 31.

Hands-on Science

Coordinating Activity

- ✂ **Breathing Flame** – Have the students see how fire requires oxygen. You will need a candle, a match, and a glass jar. (**Adults Only**) Use the match to light the candle, and have the students observe the flame. Then, turn the jar upside down, and place it over the candle. Have the students observe what happens. (The students should see the flame burn for a bit and then go out. This is because the oxygen that is trapped inside of the jar is used up by the flame. Once it is gone, the flame goes out.)

✂ **(Optional) Smelly Sulfur** – Have the students Have the students experience the smell of sulfur! You will need a vegetable with a high sulfur content, such as cabbage, brussels sprouts, or turnips. Cut the vegetable in half, smell both halves, and set one half aside. Then, cut the remaining half in half again and boil the two quarters. After 10 minutes remove one of the quarters and have the students compare the smells between the cooked and uncooked portions. Wait 10 more minutes and then remove the remaining quarter from the boiling water. Have the students compare the smells of all three slices. (The students should note an increasing "rotten-eggs" smell in the cooked portions of the vegetable. The heat causes the bonds to break, which releases the stinky sulfur-containing compound, hydrogen sulfide.)

To-Do List

Essential

- ☐ Read the appropriate reading assignment.
- ☐ Define oxidation.
- ☐ Complete the lapbook or notebook assignments.
- ☐ Do the "Breathing Flame" activity.

Optional

- ☐ Get one or more of the library books to read.
- ☐ Fill out a lab report sheet for one of the activities.
- ☐ Do the "Smelly Sulfur" activity.

Lesson 9: Halogens

Information

Reading Assignments

- 🕮 **Younger Students** – *Basher Science Periodic Table* p. 116 Halogens, p. 118 Fluorine, p. 124 Iodine
- 🕮 **Older Students** – *DK Eyewitness The Elements* pp. 62-65 Halogens

(Optional) Books from the Library

- 🕮 *Fluorine (Understanding the Elements of the Periodic Table)* by Heather Hasan
- 🕮 *The Elements: Iodine* by Leon Gray
- 🕮 *Iodine (Understanding the Elements of the Periodic Table)* by Kristi Lew

Notebooking

Vocabulary

Go over the following word with your students. Then, have them create flashcard or copy the definition into the glossary.

- **Ion** – An atom or group of atoms that has become charged by gaining or losing one or more electrons. (Flashcard LT p. 67; Glossary NB p. 57)

Mini-book Assembly Instructions

- **Lapbook** – Have the students work on the Central Periodic Table. This week, have the students color the halogens purple.
- **Lapbook** – Have the students work on the Halogens Flip-book on LT p. 57. Have the students cut along the solid lines and color the group on the cover purple. Next, have the students write several characteristics of halogens on the characteristics tab, as well as several facts about fluorine and iodine on the respective tabs. Then, have them line the pages up and staple the sheets together. Finally, glue the flip-book into the lapbook.
- **Notebook** – Have the students dictate, copy, or write one to four sentences about halogens, fluorine, and iodine on NB p. 32.

Hands-on Science

Coordinating Activity

- ✂ **Fluoride Paste** – Fluorine is often used in toothpaste to help protect the enamel on teeth from being dissolved. Have the students test this ability using an egg. You will need two eggs, two cups, fluoride toothpaste, plastic wrap, and white vinegar. Coat one of the eggs with the toothpaste, wrap it in plastic wrap, and set it in the fridge overnight. After 24 hours, gently rinse off any excess toothpaste with warm water and mark it with an "F" using a permanent maker. Then, set both eggs in a cup, cover them with vinegar, and watch what happens. (The students should see that the egg marked with an "F" does not dissolve nearly as quickly as the one without.)

✂ **(Optional) Secret Message** – Have the students use iodine to reveal a secret message. You will need a shallow pan, tincture of iodine, water, paper, Q-tip, lemon juice, and a cup. Pour some lemon juice into a cup. Then, have the students use the lemon juice and Q-tip to write a message on the paper. Meanwhile, mix together a cup of water and 20 drops of tincture of iodine in the bowl until well mixed. Once the juice has dried on the paper, have the students give you the sheet. Place it into the iodine solution in the pan to reveal the message!

To-Do List

Essential

- ☐ Read the appropriate reading assignment.
- ☐ Define ion.
- ☐ Complete the lapbook or notebook assignments.
- ☐ Do the "Fluoride Paste" activity.

Optional

- ☐ Get one or more of the library books to read.
- ☐ Fill out a lab report sheet for one of the activities.
- ☐ Do the "Secret Message" activity.

Lesson 10: Noble Gases

Information

Reading Assignments

- **Younger Students** – *Basher Science Periodic Table* p. 128 Noble Gases, p. 192 Helium, p. 193 Neon
- **Older Students** – *DK Eyewitness The Elements* pp. 66-69 Noble Gases

(Optional) Books from the Library

- *Hydrogen and the Noble Gases (True Books: Elements)* by Salvatore Tocci
- *Krypton (Understanding the Elements of the Periodic Table)* by Janey Levy

Notebooking

Vocabulary

Go over the following word with your students. Then, have them create flashcard or copy the definition into the glossary.

- **Inert** – An element that is completely nonreactive. (Flashcard LT p. 67; Glossary NB p. 56)

Mini-book Assembly Instructions

- **Lapbook** – Have the students work on the Central Periodic Table. This week, have the students color the noble gases gray.
- **Lapbook** – Have the students work on the Noble Gases Flip-book on LT p. 58. Have the students cut along the solid lines and color the group on the cover gray. Next, have the students write several characteristics of noble gases on the characteristics tab as well as several facts about helium and neon on the respective tabs. Then, line the pages up and staple the sheets together. Finally, glue the flip-book into the lapbook.
- **Notebook** – Have the students dictate, copy, or write one to four sentences about noble gases, helium, and neon on NB p. 33.

Hands-on Science

Coordinating Activity

- **Noble Change** – Have the students observe how a noble gas can change the sound of your voice. You will need a helium-filled balloon and scissors. (**CAUTION** – Helium is non-toxic, but it can cause light-headedness. Do NOT let the students do this activity – you need to demonstrate it for them. Do NOT repeat it more than once and be sure to follow the directions.) Pinch the base of a helium-filled balloon just under the knot and cut off the end so that you can get some of the gas to release. Exhale completely and place the cut balloon just under your chin. As you inhale, open the balloon so that you breathe in some of the helium. Then, start to talk or sing so that the students can observe the difference in your voice. (Your students should observe that your voice now has a squeaky quality to it and may even seem to be a bit higher. This is because Helium is six times lighter than air, so the sound waves of your voice speed up as they pass through the gas, causing your voice to sound squeaky.)

- **(Optional) Helium's History** – Have the students learn more about the history of helium-filled blimps by watching the following video:
    - https://www.youtube.com/watch?v=vfpv6JXMaGM

To-Do List

Essential

- ☐ Read the appropriate reading assignment.
- ☐ Define inert.
- ☐ Complete the lapbook or notebook assignments.
- ☐ Do the "Noble Change" activity.

Optional

- ☐ Get one or more of the library books to read.
- ☐ Fill out a lab report sheet for one of the activities.
- ☐ Do the "Helium's History" activity.

Lesson 11: Lanthanides

Information

Reading Assignments

- **Younger Students** – *Basher Science Periodic Table* p. 140 Lanthanides, p. 142 Lanthanum, p. 146 Neodymium
- **Older Students** – *DK Eyewitness The Elements* pp. 36-39 Lanthanides

(Optional) Books from the Library

- *The Lanthanides (Elements)* by Richard Beatty

Notebooking

Vocabulary

Go over the following word with your students. Then, have them create flashcard or copy the definition into the glossary.

- **Refraction** – The bending of light as it passes through a different medium. (Flashcard LT p. 67; Glossary NB p. 62)

Mini-book Assembly Instructions

- **Lapbook** – Have the students work on the Central Periodic Table. This week, have the students color the lanthanides brown.
- **Lapbook** – Have the students work on the Lanthanides Flip-book on LT p. 59. Have the students cut along the solid lines and color the group on the cover brown. Next, have the students write several characteristics of lanthanides on the characteristics tabs, as well as several facts about lanthanum and neodymium on the respective tabs. Then, line the pages up and staple the sheets together. Finally, glue the flip-book into the lapbook.
- **Notebook** – Have the students dictate, copy, or write one to four sentences about lanthanides, lanthanum, and neodymium on NB p. 34.

Hands-on Science

Coordinating Activity

- **Moving Pencils** – Have the students see how refraction works in liquids. You will need the three cups, three pencils, and three different clear liquids (i.e., water, alcohol, and corn syrup). Fill each of the three cups halfway with one of the clear liquids. Have the students place a pencil in each of the cups. Then, have them observe what has happened to the pencils. (The students should have seen that the pencils appeared to shift or move in the different liquids. This is due to refraction, or the bending of light. One of the Lanthanide elements,

Lanthanum, is often used in quality glass lenses so that the image we see is closer to the original one.)

✂ **(Optional) Neodymium Magnets** – Have the students play with neodymium magnets! Be careful to supervise them well with this activity as neodymium magnets are very strong and very dangerous if swallowed.

To-Do List

Essential

- ☐ Read the appropriate reading assignment.
- ☐ Define refraction.
- ☐ Complete the lapbook or notebook assignments.
- ☐ Do the "Moving Pencils" activity.

Optional

- ☐ Get one or more of the library books to read.
- ☐ Fill out a lab report sheet for one of the activities.
- ☐ Do the "Neodymium Magnets" activity.

Lesson 12: Actinides

Information

Reading Assignments

- **Younger Students** – *Basher Science Periodic Table* p. 158 Actinides, p. 163 Uranium, p. 167 Americium
- **Older Students** – *DK Eyewitness The Elements* pp. 40-45 Actinides

(Optional) Books from the Library

- *Radioactive Elements* by Tom Jackson
- *The 15 Lanthanides and the 15 Actinides (Understanding the Elements of the Periodic Table)*by Kristi Lew

Notebooking

Vocabulary

Go over the following word with your students. Then, have them create flashcard or copy the definition into the glossary.

- **Radioactive Decay** – The process by which a nucleus ejects particles through radiation, becoming the nucleus of a series of different elements until stability is reached. (Flashcard LT p. 68; Glossary NB p. 61)

Mini-book Assembly Instructions

- **Lapbook** – Have the students work on the Central Periodic Table. This week, have the students color the actinides white.
- **Lapbook** – Have the students work on the Actinides Flip-book on LT p. 60. Have the students cut along the solid lines and color the group on the cover white. Next, have the students write several characteristics of actinides on the characteristics tabs, as well as several facts about uranium and americium on the respective tab. Then, line the pages up and staple the sheets together. Finally, glue the flip-book into the lapbook.
- **Notebook** – Have the students dictate, copy, or write one to four sentences about actinides on NB p. 35.

Hands-on Science

Coordinating Activity

- **Candy Half-Life** – Have the students learn about a radioactive half-life using candy, or another bite-sized food. You will need bite-sized candy or food, such as raisins or cereal puffs, and a timer. Give the students 32 pieces of the bite-sized food. After 2 minutes, have them eat 16 pieces. After 2 more minutes, have them eat 8 pieces. After 2 more minutes, have them eat 4 pieces. After 2 more minutes, have them eat 2 pieces. After 2 more minutes, have them eat 1 piece. After 2 more minutes, have them break the 1 piece in half and eat one of the halves. After 2 more minutes, have the students eat any of the remaining crumbs. (This demonstration was meant to give the students a mental picture of how a half-life works. Many of the actinide elements are radioactive, which means that they are unstable so the elements decay with a half-life. The half-life period depends on the element and how radioactive it is.)

Review Sheet

Periodic Table Review Sheet Answers

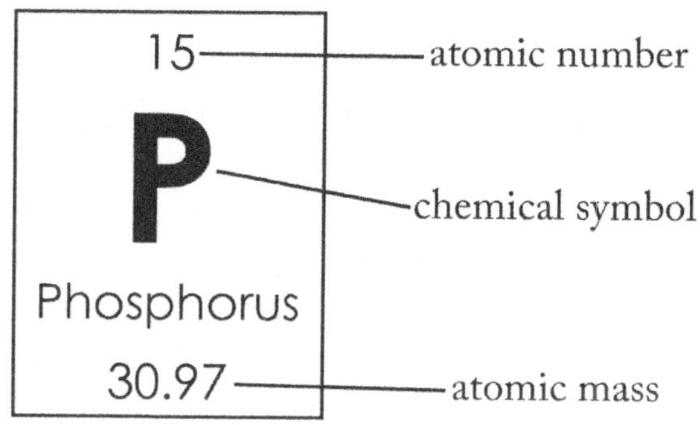

1. See the labeled image-->
2. True
3. False (An element is made up of one single type of atom.)
4. True
5. Gas
6. Soft metals; Often bonded to oxygen; React easily
7. False (Magnesium is a solid metal that burns with a bright white light.)
8. Bond
9. True
10. True
11. False (Boron is the only nonmetal in the group of metallic boron elements.)
12. Students' answers will vary, but they should include that carbon is the basis for all life.
13. True
14. 80 %
15. Important
16. Students' answers will vary, but they should include that oxygen is fuel for our body or that it is the powerhouse behind most chemical reactions on Earth.
17. False (The elements in the halogens group are very reactive.)
18. Least
19. False (Helium is lighter than air.)
20. True
21. True

To-Do List

Essential

- ☐ Read the appropriate reading assignment.
- ☐ Define radioactive decay.
- ☐ Complete the lapbook or notebook assignments.
- ☐ Do the "Candy Half-Life" activity.

Optional

- ☐ Get one or more of the library books to read.
- ☐ Fill out a lab report sheet for one of the activities.
- ☐ Complete the Periodic Table Review Sheet. (pp. 124-125)

SCIENCE CHUNKS

Curie Unit

Lesson 1: Marie Curie

Information

Reading Assignments

📖 **All Students** – Prologue: "Who Was Marie Curie?" to Chapter 4: "Two Loves" from *Who Was Marie Curie?* by Megan Stine
Note—You can read these chapters all in one sitting or break them up into two sessions as follows:
- Day 1 – Prologue: "Who Was Marie Curie?" and Chapter 1: "Eager to Learn"
- Day 2 – Chapter 2: "The Secret School" to Chapter 4: "Two Loves"

Discussion Questions

When you are done reading, you can ask the students the following questions:

Prologue: "Who Was Marie Curie?"

- What did Marie Curie get a Nobel Prize for? (Marie Curie got a Nobel Prize for her work in discovering radium.)

Chapter 1: "Eager to Learn"

- What was Maria's life like when she was growing up? (Maria was the youngest child, with three sisters and a brother. Her mother was a teacher and a principal. Her father was a science teacher and eventually ran his own school. They lived in Poland when it was under Russia's control, but they did not like the Russians. She had a hard life; her mother was sick and eventually died. Her house was full of boys from her father's school, and her sister died when she was young. But she was very smart and even graduated high school early.)
- **(Older Students)** How do you think Marie was affected by the fact that both of her parents were teachers? (Answers will vary.)

Chapter 2: "The Secret School"

- How did Maria and her sister continue their education? (They both attended a secret college in Poland called the Flying University. Then, Maria agreed to stay behind and work so that her sister, Bronia, could go to the Sorbonne in Paris. Then, when Bronia was done, she invited Maria to stay with her while Maria attended the Sorbonne.)

Chapter 3: "Hungry but Happy"

- What was college life like for Marie? (Marie loved the Sorbonne and did very well, graduating top of her class. There were very few women in her classes, and she was very poor during that time. She had to endure a lot of hardships, like not having heat or enough food.)
- **(Older Students)** Why do you think Marie was willing to endure so much to get her education? (Answers will vary.)

Chapter 4: "Two Loves"

- How did Marie end up staying in Paris, France? (At first, Marie stayed to finish a math degree. Then, she got a job in a lab studying magnetism and steel. After a brief return to Warsaw, she came back to Paris and married Pierre Curie. They stayed there to work in the sciences together.)

(Optional) Books from the Library

- *DK Biography: Marie Curie* by DK Publishing
- *Marie Curie (Giants of Science)* by Kathleen Krull
- *What Makes a Magnet? (Let's-Read-and-Find-Out Science 2)* by Franklyn M. Branley and True Kelley

Notebooking

Vocabulary

Go over the following words with your students. Then, have them create flashcards or copy the definitions into the glossary.

- **Magnet** – An object that attracts iron, steel, and metals. (Flashcard LT p. 75; Glossary NB p. 58)
- **Magnetism** – The force that causes magnets and metals to stick together. (Flashcard LT p. 76; Glossary NB p. 58)

Writing Instructions

- **Lapbook** – Have the students work on the Marie Curie tab-book on LT p. 72. Have them cut out the mini-book pages and color the cover. Next, have the students tell you what they have learned about Marie Curie and her work on the respective tabs. Then, have them set the pages aside for next week.

- **Lapbook** – Have the students complete the Curie timeline sheet on LT p. 74. Have them cut out the timeline sheet and color the pictures. Next, have the students add one to three of the significant dates (see below) to the timeline.

- **Lapbook** – Have the students cut out and glue the vocabulary pocket on LT p. 75 into the lapbook.

- **Notebook** – Have the students add the following dates to the Curie timeline notebooking page on NB p. 40.
  - November 7th, 1867 – Maria Sklodowska is born in Warsaw.
  - 1878 – Maria's mother dies.
  - 1883 – Maria graduates from high school.
  - 1891 – Maria goes to Paris to study at the Sorbonne, taking on the French version of her name, Marie.
  - July 26th, 1895 – Marie and Pierre Curie get married.
  - 1903 – Marie wins the Nobel Prize.

Notebook – Have the students dictate, copy, or write information about Marie Curie and her work on the Curie notebook pages on NB pp. 38-39.

Hands-on Science

Coordinating Activity

- **Magnetic Slime** – In Chapter 4, we learned that Marie's first job in the lab involved studying magnetism. In honor of that, have the students make some magnetic slime and use a few magnets to play with it. You will need white glue, water, iron filings, Borax, and magnets. In a plastic baggie, mix 4oz of white glue, 4oz of of water, and about a half teaspoon of iron filings. In a cup, mix a quarter cup of water and half a teaspoon of Borax. Then, add the Borax solution to the baggie, and massage the bag for a few minutes until a nice firm slime has formed. Then, the magnetic fun begins! Have the students use the magnets to play around with the slime and pull it in different directions. They can also test the strength of a magnet by seeing how much slime it can pull up.
- **(Optional) Hot Ice** – This experiment is a nod to Marie's love of playing with science. It is very cool when it works, but it's a bit temperamental. You will need vinegar and baking soda for the experiment. Directions and troubleshooting tips can be found at:
  - http://chemistry.about.com/od/homeexperiments/a/make-hot-ice-sodium-acetate.htm

  Or you can watch a video of this experiment here:
  - https://www.youtube.com/watch?v=pzHiVGeevZE

To-Do List

Essential

- ☐ Read the appropriate reading assignment.
- ☐ Define magnet and magnetism.
- ☐ Do the "Magnetic Slime" activity.
- ☐ Complete the lapbook or notebook assignments.

Optional

- ☐ Get one or more of the library books to read.
- ☐ Do the "Hot Ice" activity.

Lesson 2: Marie Curie

Information

Reading Assignments

📖 **All Students** – Chapter 5: "Marie's Discovery" to Chapter 9: "A Family of Scientists" from *Who Was Marie Curie?* by Megan Stine
Note—You can read these chapters all in one sitting or break them up into two sessions as follows:

- Day 1 – Chapter 5: "Marie's Discovery" and Chapter 7: "Fame and Fortune"
- Day 2 – Chapter 8: "Misery" to Chapter 9: "A Family of Scientists"

Discussion Questions

When you are done, you can ask the students the following questions:

Chapter 5: "Marie's Discovery"

- What did Marie discover about pitchblende? (Marie discovered that pitchblende gave off more rays than uranium alone, meaning that there must be another radioactive element in it.)
- What happened when Marie tried to isolate polonium? (When Marie tried to isolate polonium from the pitchblende, she discovered another radioactive element, which she named radium.)

Chapter 6: "It Glows!"

- What did Marie and Pierre do to isolate radium? (Marie and Pierre spent three years isolating radium. They had to sift, boil, and separate tons of crushed pitchblende.)
- What was special about the radium salts the Curies produced? (The radium salts glowed in the dark, but they also caused damage to the Curies' skin and insides.)
- **(Older Students)** Why do you think Marie and Pierre were willing to endure the opposition and difficulties they faced to make their discovery? (Answers will vary.)

Chapter 7: "Fame and Fortune"

- What happened to Marie and Pierre after they won the Nobel Prize? (After they won the Nobel Prize, Pierre was given a professorship at the Sorbonne, and he and Marie were given a better lab to work in.)
- **(Older Students)** Why was there such a radium craze in the early 1900s? (Students should share their own thoughts, but their answers should include some of the following information: People were so in love with radium because it not only glowed in the dark but was also thought to cure diseases. They had no idea that radium was actually what was making them sick.)

Chapter 8: "Misery"

- What was Marie's life like after Pierre died? (After Pierre died, Marie was very sad, but eventually, she was invited to take over his teaching job at the Sorbonne. After that, she moved to the country and created a school of her own in her home. She also won a second Nobel Prize.)

Chapter 9: "A Family of Scientists"

- How did Marie Curie help out during World War I? (During World War I, Marie Curie hid all the radium in France from the Germans, and she also created a portable X-ray machine to help soldiers.)
- What did Marie Curie's children do to carry on her legacy? (To carry on her mother's legacy, Irene took over the Radium Institute and continued to work in the sciences, eventually earning her own Nobel Prize. Marie's other daughter Eve wrote a book about her life.)

(Optional) Books from the Library

- *Marie Curie's Search for Radium* by Beverly Birch
- *Marie Curie: The Woman Who Changed the Course of Science* by Philip Steele
- *Radioactive Elements* by Tom Jackson

Notebooking

Vocabulary

Go over the following words with your students. Then, have them create flashcards or copy the definitions into the glossary.

- **Radioactivity** – The release of radiation from the nuclei of unstable atoms. (Flashcard LT p. 76; Glossary NB p. 61)

Writing Instructions

- **Lapbook** – Have the students work on the Marie Curie tab-book on LT pp. 72-73. Have them cut out the mini-book pages and color the cover. Next, have the students tell you what they have learned about Marie Curie and her work on the respective tabs. Then, have them staple the pages of the booklet together and glue the pages into the lapbook.
- **Lapbook** – Have the students complete the Curie timeline sheet on LT p. 74. Have them cut out the timeline sheet and color the pictures. Next, have the students add one to three of the significant dates (see below) to the timeline.
- **Notebook** – Have the students add the following dates to the Curie timeline notebooking page on NB p. 40.
  - September 12th, 1897 – Marie and Pierre Curie have a baby girl and name her Irene.
  - 1898 – Marie discovers both polonium and radium in a piece of pitchblende.
  - July 1902 – Marie isolates a few grains of pure radium.
  - 1903 – Marie gets her PhD from the Sorbonne.
  - December 6th, 1904 – Marie and Pierre have another child, a girl named Eve.
  - April 1906 – Pierre dies.
  - 1911 – Marie is awarded a second Nobel Prize.

- 1914 – The Radium Institute, later known as the Curie Institute, is built.
- 1934 – Marie Curie dies.

Notebook – Have the students dictate, copy, or write information about Marie Curie and her work on the Curie notebook pages on NB p. 39.

Hands-on Science

Coordinating Activity

Science Cookies – We have learned that Maria's love of science began early thanks to the games and lessons her father provided and that she spent her life pursuing science. In honor of that, have the students create a super fun chemical reaction, one that may or may not have gotten Marie Curie interested in science. You will need a cup, a plate, food coloring, an eye dropper, a cookie cutter, baking soda, and white vinegar. Mix several drops of food coloring with the vinegar in the cup. Then, place the cookie cutter on the plate, and fill it halfway with baking soda. Now, use the eyedropper to squirt colored vinegar onto the baking soda! (You should see lots of fizzing where the vinegar meets the baking soda. This is due to the gas carbon dioxide, which is released when an acid, like vinegar, reacts with a base, like baking soda.)

(Optional) Nuclear Power Plant – Have the students learn more about nuclear power by watching the following video:

https://www.youtube.com/watch?v=d7LO8lL4Ai4

Note—This video does touch on the dangers of nuclear power vs. other power methods. Please preview this video to make sure it is appropriate for your child.

Review Sheet

Curie Review Sheet Answers

1. B, C, A
2. True
3. False (Marie Curie spent three years and a whole lot of work isolating just a few grains of radium.)
4. True
5. Two
6. Answers will vary.

To-Do List

Essential

- ☐ Read the appropriate reading assignment.
- ☐ Define radioactivity.
- ☐ Do the "Science Cookies" activity.
- ☐ Complete the lapbook or notebook assignments.

Optional

- ☐ Get one or more of the library books to read.
- ☐ Do the "Nuclear Power Plant" activity.
- ☐ Complete the Curie Review Sheet on p. 127.

SCIENCE CHUNKS

Force Unit

Lesson 1: Forces

Information

Reading Assignments

- **Younger Students** – *Basher Science Physics* p. 20 Force
- **Older Students** – *Usborne Science Encyclopedia* pp. 118-119 Forces

(Optional) Books from the Library

- *Forces Make Things Move (Let's-Read-and-Find-Out Science 2)* by Kimberly Bradley and Paul Meisel
- *Forces (Science Readers)* by Debra J. Housel
- *Push and Pull (Rookie Read-About Science)* by Patricia J. Murphy
- *Pushes and Pulls (TIME FOR KIDS® Nonfiction Readers)* by Sharon Coan

Notebooking

Vocabulary

Go over the following words with your students. Then, have them create flashcards or copy the definitions into the glossary.

- **Force** – A push or a pull on an object. (Flashcard LT p. 86; Glossary NB p. 54)

Writing Instructions

- **Lapbook** – Have the students begin the Forces and Motion lapbook by cutting out and coloring the cover on LT p. 79. Then, have the students glue the sheet onto the front.
- **Lapbook** – Have the students work on the Force Tab-book on LT p. 81. Have them cut out the force page and the cover page for the tab-book. Then, have the students color the pictures and add a sentence of two about forces on the force page. Have them put the pages in a safe place because they will be adding to this tab-book throughout the unit.
- **Notebook** – Have the students dictate, copy, or write three to five sentences on forces on the forces notebooking page on NB p. 42.

Hands-on Science

Coordinating Activity

- **Forces Lab** – Have the students do a quick forces lab. You will need a toy car, string, tape, several books, and a cardboard sheet. Have the students perform each of the three tests on the toy car:
    - **Test #1 (Push):** Set the car on a flat surface, and observe what happens. Then, have the

student gently push the back of the car and observe what happens.

- **Test #2 (Pull):** Use the tape to attach a piece of string to the front of the car. Then, have the student gently pull on the string and observe what happens.
- **Test #3 (Both):** On a smooth surface, stack several books on top of each other, and then set the cardboard up against the books and the surface, creating a ramp. Hold the car at the top of the ramp, and gently push it down the ramp, observing what happens.

The students should see that in each test, the car did not move until they either pushed or pulled the car.

(Optional) Force Race – Have the students compete to see who can force a balloon over a line first You will need an empty soda bottle and a balloon for each player. Begin by drawing two lines at either end of a room or outside. Then, give each player an empty soda bottle and a balloon. Have them line up on the first line. When you yell "go," the players will use their empty soda bottle to push their balloon toward the second line. The player to cross the line first wins the race.

To-Do List

Essential

☐ Read the appropriate reading assignment.

☐ Define force.

☐ Complete the lapbook or notebook assignments.

☐ Do the "Forces Lab" activity.

Optional

☐ Get one or more of the library books to read.

☐ Fill out a lab report sheet (p. 103) for one of the activities.

☐ Do the "Force Race" activity.

Lesson 2: Balance

Information

Reading Assignments

- 🕮 **Younger Students** – *Article: "Balance"* p. 47
- 🕮 **Older Students** – *Usborne Science Encyclopedia* p. 120 Combining Forces and Equilibrium

(Optional) Books from the Library

- 🕮 *Balances (Science Tools)* by Adele D. Richardson
- 🕮 *Make it Balance (Let's Explore Science)* by Claudette Williams
- 🕮 *Balance (First Step Nonfiction Simple Tools)* by Sheila Rivera

Notebooking

Vocabulary

Go over the following words with your students. Then, have them create flashcards or copy the definitions into the glossary.

- **Balance** – The point at which all the forces acting on an object cancel each other out. (Flashcard LT p. 87; Glossary NB p. 53)
- **Resultant force** – The combined effect of the forces acting on an object. (Flashcard LT p. 87; Glossary NB p. 62)

Writing Instructions

- **Lapbook** – Have the students complete the Balance Mini-book on LT p. 84. Have them cut out and fold the template. Have the students color the pictures on the cover. Then, have the students add a sentence or two about balance on the inside. Finally, have them glue the mini-book into the lapbook.

- **Notebook** – Have the students dictate, copy, or write three to five sentences on balance on the balance notebooking page on NB p. 43.

Hands-on Science

Coordinating Activity

- **Balance Tower** – Have the students do a balance tower. You will need a bag of gumdrops, toothpicks, and a plate. Begin by making several shapes out of gumdrops and toothpicks, such as squares, triangles, and octagons. Have the students test the strength of the different shapes by standing the shapes up and pressing on different spots. Have them build a gumdrop-toothpick tower that stands tall using what they have learned from the first two steps. Once built, have the students place a toothpick and gumdrop in random places until the tower topples over. The purpose of this activity was for the students to see balance in action.

✂ **(Optional) Balancing Robot** – Have the students make a balanced robot. You will need cardstock, scissors, two pennies, poster putty, and markers, crayons, or colored pencils. The directions and a free robot printable for this project can be found here:

- https://buggyandbuddy.com/science-kids-balancing-robot-free-printable/

To-Do List

Essential

- ☐ Read the appropriate reading assignment.
- ☐ Define balance and resultant force.
- ☐ Complete the lapbook or notebook assignments.
- ☐ Do the "Balance Tower" activity.

Optional

- ☐ Get one or more of the library books to read.
- ☐ Fill out a lab report sheet for one of the activities.
- ☐ Do the "Balancing Robot" activity.

Lesson 3: Gravity

Information

Reading Assignments

- **Younger Students** – *Basher Science Physics* p. 26 Gravity, p. 10 Mass, p. 12 Weight
- **Older Students** – *Usborne Science Encyclopedia* pp. 130-131 Gravity

(Optional) Books from the Library

- *Gravity Is a Mystery (Let's-Read-and-Find-Out Science 2)* by Dr. Franklyn M. Branley and Edward Miller
- *Gravity (Blastoff! Readers: First Science)* by Kay Manolis
- *Gravity (First Step Nonfiction Forces and Motion)* by Robin Nelson
- *What Is Gravity? (Rookie Read-About Science)* by Lisa Trumbauer

Notebooking

Vocabulary

Go over the following words with your students. Then, have them create flashcards or copy the definitions into the glossary.

- **Gravity** – The pulling force that attracts objects to each other. (Flashcard LT p. 88; Glossary NB p. 55)
- **Mass** – The amount of matter contained in an object. (Flashcard LT p. 88; Glossary NB p. 58)
- **Weight** – A measure of the strength of the pull of gravity on an object. (Flashcard LT p. 89; Glossary NB p. 64)

Writing Instructions

- **Lapbook** – Have the students work on the Force Tab-book on LT p. 82. Have them cut out the gravity page and color the pictures. Then, have the students add a sentence or two about gravity, mass, and weight on the gravity page. Have them put the pages in a safe place because they will be adding to this tab-book throughout the unit.
- **Notebook** – Have the students dictate, copy, or write three to five sentences on gravity on the gravity notebooking page on NB p. 64.

Hands-on Science

Coordinating Activity

- **Rolling Pencil** – Have the students do a rolling pencil activity. You will need a pencil, a cardboard sheet (or a hardcover book), and a flat surface. Place the pencil on top of the cardboard sheet (or hardcover book) on the flat surface, and have the students observe if the pencil is moving and write this down on the chart on the activity sheet on p. 102. *(It should not be.)* Have the students pick up the end of the cardboard sheet (or hardcover book) nearest to the pencil and lift it until the sheet (or book) is at an angle of about 45° to the surface. Have the students observe what happens to the pencil and write this down on the chart on the activity sheet. *(It should roll off and onto the flat surface.)* Have the students attempt to roll the pencil back up the cardboard sheet (or book) to where the pencil was located before. Have them observe what happens and write this down on the chart on the activity sheet. *(It should roll up a bit and then fall back down to the flat surface.)* The purpose of this activity was for the students to see the force of gravity in action.

✂ **(Optional) Gravity Painting** – Have the students use gravity to create a masterpiece. You will need several pom-poms, paper, and paint. Add a bit of water to the paint to thin it out a bit. Then lay out a piece of paper on the floor. Have the students dip a pom-pom in paint and then drop it from shoulder height onto the paper.

To-Do List

Essential

- ☐ Read the appropriate reading assignment.
- ☐ Define gravity, mass, and weight.
- ☐ Complete the lapbook or notebook assignments.
- ☐ Do the "Rolling Pencil" activity.

Optional

- ☐ Get one or more of the library books to read.
- ☐ Fill out a lab report sheet for one of the activities.
- ☐ Do the "Gravity Painting" activity.

Lesson 4: Friction

Information

Reading Assignments

- **Younger Students** – *Basher Science Physics* p. 24 Friction
- **Older Students** – *Usborne Science Encyclopedia* pp. 124-125 Friction

(Optional) Books from the Library

- *Friction (True Books: Physical Science)* by Matt Mullins
- *Friction (Science Readers: Content and Literacy)* by Suzanne I. Barchers
- *What Is Friction? (Rookie Read-About Science)* by Lisa Trumbauer and David Larwa
- *Why Do Moving Objects Slow Down?: A Look at Friction* by Jennifer Boothroyd

Notebooking

Vocabulary

Go over the following words with your students. Then, have them create flashcards or copy the definitions into the glossary.

- **Friction** – The force that causes moving objects to slow down when they are touching. (Flashcard LT p. 89; Glossary NB p. 55)
- **Lubricant** – A substance used to reduce friction. (Flashcard LT p. 90; Glossary NB p. 57)

Writing Instructions

- **Lapbook** – Have the students work on the Force Tab-book on LT p. 82. Have them cut out the friction page and color the pictures. Then, have the students add a sentenceof two about friction and lubricants on the friction page. Have them put the pages in a safe place because they will be adding to this tab-book throughout the unit.
- **Notebook** – Have the students dictate, copy, or write three to five sentences on friction on the friction notebooking page on NB p. 45.

Hands-on Science

Coordinating Activity

- **Friction Roll** – Have the students do a friction roll activity. You will need a marble, a cookie sheet (or smooth cutting board), a paper, felt, and a book. Use the book to prop up one end of the cookie sheet (or cutting board) so that the marble will roll from one end to the other. (*You don't want too much of an incline because you don't want the marble to go too fast, so pick a book that is less than an inch thick.*) Have the students set the marble at the top of the cookie sheet (or cutting board) and let go. Have them count how long it takes for the marble to reach the bottom and record that on their activity sheet. Next, place the paper in the cookie sheet (or cutting board) so that it covers the surface. Once again, have the students set the marble at the top and let go. Have them count how long it takes for the marble to reach the bottom and record that on their activity sheet. Then, place the felt in the cookie sheet (or cutting board) so that it covers the surface. Once more, have the students set the marble at the top and let go. Have them count how long it takes for the marble to reach the bottom and record that on their activity sheet. The students should see that the paper and felt both slowed the marble down.

✂ **(Optional) Friction Block** – Have the students examine how different types of materials create more or less friction. You will need a 3-foot section of a 2x4 board, a 1x2x4 block (average-sized wooden block), a sheet of sandpaper, a washcloth, several large books, and a timer. Begin by propping up the board using the books so that the top is 18 inches higher than the bottom. Then, send the block down the board, recording the time it takes to get to the bottom. Next, cover the block with the washcloth, and send it down the board, recording the time it takes to get to the bottom. Finally, cover the block with the sandpaper, and send it down the board, recording the time it takes to get to the bottom. (*The students should see that it took longer for the block to go down the ramp on the second and third times, which is due to the friction caused by the washcloth and sandpaper. You can also have the students test out several lubricants to see if they speed up the block descent.*)

To-Do List

Essential

- ☐ Read the appropriate reading assignment.
- ☐ Define friction and lubricant.
- ☐ Complete the lapbook or notebook assignments.
- ☐ Do the "Friction Roll" activity.

Optional

- ☐ Get one or more of the library books to read.
- ☐ Fill out a lab report sheet for one of the activities.
- ☐ Do the "Friction Block" activity.

Lesson 5: Floating

Information

Reading Assignments

- **Younger Students** – *Basher Science Physics* p. 14 Density
- **Older Students** – *Usborne Science Encyclopedia* pp. 138-139 Floating

(Optional) Books from the Library

- *What Floats? What Sinks?: A Look at Density* by Jennifer Boothroyd
- *Dive! Dive! Dive!: Buoyancy (Raintree Fusion: Physical Science)* by Isabel Thoma

Notebooking

Vocabulary

Go over the following words with your students. Then, have them create flashcards or copy the definitions into the glossary.

- **Density** – A measure of the amount of matter (mass) in a substance compared to its volume. (Flashcard LT p. 82; Glossary NB p. 53)

Writing Instructions

- **Lapbook** – Have the students complete the Force Tab-book on LT p. 75. Have them cut out the density page and color the pictures. Then, have the students add a sentence or two about density on the density page. Once they are done, have them staple all the pages for the tab-book together and glue it into the lapbook.
- **Lapbook** – Have the students complete the Floating Mini-book on LT p. 77. Have them cut out and fold the template. Have the students color the pictures on the cover. Then, have the students add a sentence or two about floating on the inside. Finally, have them glue the mini-book into the lapbook.
- **Notebook** – Have the students dictate, copy, or write three to five sentences on floating on the floating notebooking page on NB p. 46.

Hands-on Science

Coordinating Activity

- **Floating Ship** – Have the students make a floating ship. You will need a balloon, an old CD, a 2-L soda bottle cap, a thin nail, and glue. Use the nail to poke at least ten tiny holes in the top of your soda bottlecap. Then, have the students glue the bottom of the bottlecap over the hole in the center of the CD so that the holes face up. After the glue dries, set the CD on a table or flat surface, and blow up the balloon. Quickly fit the balloon over the soda bottlecap, and have the students observe what happensand write they see on the activity sheet on p. 102. The students should see that the simple CD ship appears to float above the surface as the ballon deflates.

✂ **(Optional) Floating Egg** – Have the students float an egg. You will need a cup, water, egg, and salt. Place an egg in a cup filled about two-thirds of the way with tap water, and observe what happens. (*The students should see the egg sink to the bottom.*) Add about a quarter to a half a cup of salt, depending on the size of your glass. Stir gently and observe what happens to the egg. (*The students should see the egg float. If it does not, you may need to add a bit more salt. The egg floats in the salt water because the liquid is more dense, which means that the egg weighs less than the salt water it displaces.*)

Review Sheet

Force Review Sheet Answers

1. False (Forces can be visible, such as a contact force, or invisible, such as gravity.)
2. Contact
3. True
4. False (When an object is at rest, there are still forces acting on it.)
5. Center of gravity
6. False (Mass is the amount of matter in an object, wheras weight is the measure of t[illegible]th of the pull of gravity on an object.)
7. False (A lubricant helps reduce friction.)
8. Air resistance (or drag)
9. The same or less
10. True

To-Do List

Essential

- ☐ Read the appropriate reading assignment.
- ☐ Define density.
- ☐ Complete the lapbook or notebook assignments.
- ☐ Do the "Floating Ship" activity.

Optional

- ☐ Get one or more of the library books to read.
- ☐ Do the "Floating Egg" activity.
- ☐ Complete the Force Review Sheet. (pp. 128-129)

SCIENCE CHUNKS

Motion Unit

Lesson 1: Dynamics

Information

Reading Assignments

- **Younger Students** – *Basher Physics* p. 22 Inertia
- **Older Students** – *Usborne Science Encyclopedia* pp. 122-123 Dynamics

(Optional) Books from the Library

- *Motion (Science Readers: Content and Literacy)* by Debra Housel
- *And Everyone Shouted, "Pull!": A First Look at Forces and Motion* by Claire Llewellyn and Simone Abel
- *Motion: Push and Pull, Fast and Slow (Amazing Science)* by Darlene R. Stille and Sheree Boyd

Notebooking

Vocabulary

Go over the following words with your students. Then, have them create flashcards or copy the definitions into the glossary.

- **Inertia** – The tendency of objects to resist a change in their movement. (Flashcard LT p. 98; Glossary NB p. 56)
- **Momentum** – A measure of an object's tendency to continue moving. (Flashcard LT p. 99; Glossary NB p. 59)

Writing Instructions

- **Lapbook** – Have the students complete the Three Laws of Motion Sheet on LT p. 94. Have them cut out the template and color the pictures. You can have the students copy the Three Laws of Motion under their respective sections or cut out and glue on the pre-written laws. Once they are done, have them glue the sheet into the lapbook.
- **Notebook** – Have the students dictate, copy, or write three to five sentences on dynamics and motion on the dynamics notebooking page on NB p. 48.

Hands-on Science

Coordinating Activities

- **Third Law** – Have the students use a balloon rocket to see the Third Law in action. You will need a straw, string (5 feet), scissors, large balloon, 2 chairs, and a tape. Have the students cut the straw in half and thread the string through the straw. As they are doing this, set the chairs up so that the backs of the chairs are facing each other about 5 feet apart. Tape or tie the ends of the string to the backs of the chairs. Have the students blow up the balloon, and twist the end so that it is easier to hold shut. Carefully tape the inflated balloon to the straw. Still holding the open end of the strawballoon rocket closed, move it to the end of the string with the open end facing one of the chair backs. Have the students let go of the open end of the straw-balloon rocket and watch what happens! The students should see the straw-balloon rocket take off along the string toward the other chair when they let go of the balloon.

(Optional) Dynamic Marbles – Have the students play a game of marbles. In the dynamic marble game, they will use one marble to hit and move another, which is a perfect example of the laws of motion! If you are not familiar with playing marbles, directions for this game can be found at the following website:

- https://www.scarymommy.com/marble-games/

To-Do List

Essential

- ☐ Read the appropriate reading assignment.
- ☐ Define inertia and momentum.
- ☐ Complete the lapbook or notebook assignments.
- ☐ Do the "Third Law" activity.

Optional

- ☐ Get one or more of the library books to read.
- ☐ Fill out a lab report sheet (p. 103) for one of the activities.
- ☐ Do the "Dynamic Marbles" activity.

Lesson 2: Speed

Information

Reading Assignments

- **Younger Students** – *Basher Physics* p. 18 Acceleration
- **Older Students** – *Usborne Science Encyclopedia* pp. 126-127 Motion

(Optional) Books from the Library

- *Motion (Blastoff! Readers: First Science)* by Kay Manolis
- *Vroom! Speed and Acceleration (TIME FOR KIDS® Nonfiction Readers)* by Stephanie Paris
- *Full Speed Ahead!: How Fast Things Go* by Cruschiform

Notebooking

Vocabulary

Go over the following words with your students. Then, have them create flashcards or copy the definitions into the glossary.

- **Acceleration** – A change in the speed or direction of an object. (Flashcard LT p. 99; Glossary NB p. 52)
- **Speed** – A measure of how fast an object is moving. (Flashcard LT p. 100; Glossary NB p. 63)

Writing Instructions

- **Lapbook** – Have the students complete the Speed versus Acceleration Shutterfold Book on LT p. 96. Have them cut the template for the shutterfold book and color the pictures. Next, have the students add a sentence about speed on the top of the inside. Then, have them add a sentence about acceleration on the bottom of the inside. Have them fold the mini-book and glue it into the lapbook.
- **Notebook** – Have the students dictate, copy, or write three to five sentences on speed and motion on the speed notebooking page on NB p. 49.

Hands-on Science

Coordinating Activities

- **Accelerate** – Have the students push a car to accelerate its motion and then let the forces of gravity and friction do the deceleration. You will need a toy car and a ramp. Have the students set the car on a flat surface and gently push the car. Have them count how long it takes to stop. Then, have them push the car down the ramp with the same force used in the previous step and count how long it takes to stop. Finally, have the students push the car up the ramp with the same force used in the previous step and count how long it takes it to stop. (**Note** –

The stoping point is the point at which it momentarily stops before heading back down the ramp.) The students should see the car accelerating and decelerating depending on the incline and direction.

(Optional) Speed Poster – Have the students make a poster depicting the differences among speed, velocity, and acceleration. Divide a piece of poster board or paper into three sections. Begin by titling the first section "Speed" and add the definition of speed along with a picture of a ball in motion. Next, title the second section "Velocity" and add the definition of velocity along with a picture of a ball in motion and a arrow showing its direction. Finally, title the third section "Acceleration" and add the definition of acceleration along with a picture of a ball being hit by a bat. For your reference, you can use the following definition for velocity on the poster:

- **Velocity** – A measure of how fast an object is moving in a given direction.

To-Do List

Essential

- ☐ Read the appropriate reading assignment.
- ☐ Define acceleration and speed.
- ☐ Complete the lapbook or notebook assignments.
- ☐ Do the "Accelerate" activity.

Optional

- ☐ Get one or more of the library books to read.
- ☐ Fill out a lab report sheet for one of the activities.
- ☐ Do the "Speed Poster" activity.

Lesson 3: Circular Motion

Information

Reading Assignments

- **Younger Students** – No pages scheduled—have the students watch the following videos to learn more about centripetal force instead:
    - Centripetal Force: https://www.youtube.com/watch?v=KvCezk9DJfk
    - What Is Centrifugal Force: https://www.youtube.com/watch?v=gRVIWWJwzfY
- **Older Students** – *Usborne Science Encyclopedia* p. 128 Motion, p. 129 Centripetal Force

(Optional) Books from the Library

- There are no additional books suggested for this week.

Notebooking

Vocabulary

Go over the following words with your students. Then, have them create flashcards or copy the definitions into the glossary.

- **Centripetal Force** – A force that keeps an object moving in a circle. (Flashcard LT p. 100; Glossary NB p. 53)
- **Torque** – The force you add to make something rotate. (Flashcard LT p. 101; Glossary NB p. 64)

Writing Instructions

- **Lapbook** – Have the students complete the Circular Motion Mini-book on LT p. 97. Have them cut out and fold the template. Have the students color the pictures on the cover. Then, have the students write the definition of centripetal force. Finally, have them glue the mini-book into the lapbook.
- **Notebook** – Have the students dictate, copy, or write three to five sentences on circular motion on the circular motion notebooking page on NB p. 50.

Hands-on Science

Coordinating Activity

- **Circular Penny** – Have the students do the circular penny activity. You will need a balloon and a penny. Have the students place the penny inside the balloon. Then, blow the balloon up, and tie it off for the students so that the penny won't come out. Now, have the students move the balloon side to side or back-and-forth to get the penny rolling along the outside of the balloon in a circular path. Then, have them stop and observe what happens to the penny.

The students should see the penny kept moving even after they stopped moving.

(Optional) Spinning Shapes – Have the students test different shapes to see how shape affects spin. You will need a sharpened pencil, thin cardboard, and scissors for this activity. Cut out several shapes from the cardboard, such as a circle, a square, and a triangle. Mark the center of each shape and use the pointed end of the scissors to poke a hole in the center of each shape big enough for the pencil to fit through. Then, let the students test the spinning capabilities of each shape by seeing which one stays up the longest.

Review Sheet

Motion Review Sheet Answers

1. Inertia, momentum
2. True
3. Greater, greater
4. All of the above
5. True
6. True
7. False (Objects moving in a circle do travel in a straight line.)

To-Do List

Essential

- ☐ Read the appropriate reading assignment.
- ☐ Define centripetal force and torque.
- ☐ Complete the lapbook or notebook assignments.
- ☐ Do the "Cicular Penny" activity.

Optional

- ☐ Get one or more of the library books to read.
- ☐ Do the "Spinning Shapes" activity.
- ☐ Complete the Motion Review Sheet. (p. 130)

SCIENCE CHUNKS

Appendix

Activity: ______________________

What I learned

Lab Report: ______

Our Tools

Our Method

What it looked like

Our Outcome

Our Insight

Is It Living or Nonliving?

All around us we can find both living and nonliving things. Biology is often called the science of life because it studies or looks at living things.

A living thing has the characteristics of life. This means that it develops, changes, and grows. All living things reproduce to make exact copies of themselves or offspring that are similar. Living things have cells, which are the tiny building blocks of life. These cells use energy to do tasks, and they need food to give them energy. Living things are also able to move as they respond and adapt to their environments. You are a living thing!

In contrast, nonliving things do not show the characteristics of life. They may appear to show one or more of the same characteristics, but they do not show them all. Icicles appear to grow and change, but they are not living. Icicles don't adapt, and they don't use energy. If it gets to hot, icicles disappear! Icicles are nonliving.

We can ask the following questions as we observe an object to know if it is living or not:

- **?** Does the object change, develop, or grow?
- **?** Does the object reproduce?
- **?** Does the object need food?
- **?** Does it move?
- **?** Does the object adapt or respond what is around it?

Mold

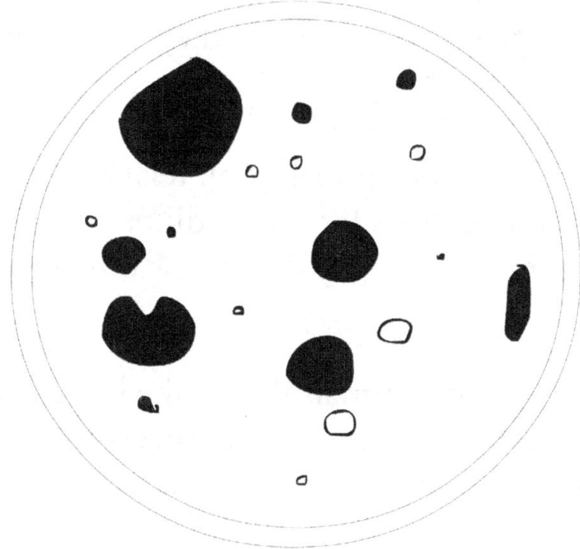

Mold is that furry or slimy stuff that grows on food that has gone bad. Sometimes it's so icky that it is hard for us to see any value in mold, but this type of fungus has a purpose!

Mold is another one of nature's decomposers. It feeds off decaying trees on the forest floor or decaying food that has been forgotten on a shelf.

Mold grows together as a colony, and the part we see is the part that makes the spores, like mushrooms.

The mold spores are released into the air and float around until they find a place to land. If conditions are right, meaning the place where the spores land is damp, or humid, the mold will grow. As the mold grows, a network of connected, thread-like structures known as mycelium form. These meet up with mycelium from other spores that have grown. A colony of mold blooms before our eyes and feeds on the decaying food. This newly formed mold colony will release its own spores, completing the mold life cycle.

Mold spores are present in every environment. Some are dangerous to humans in large quantities because they can cause allergic reactions and respiratory problems. Some mold spores produce useful chemicals.

One of the most famous molds, *Penicillium notatum*, grows on decaying fruit. This type of mold produces a powerful antibiotic that kills bacteria but doesn't always harm the human body.

Life Cycles

Animals go through a series of changes over their lifetime. They are born new and ready to learn. They grow and develop. Then, they reach adulthood, a time when they help begin the life cycle for a new generation.

Some animals look relatively the same from young to adult, like elephants. Some animals look a bit different from young to adult, like humans. And some animals look very different from young to adult, like frogs.

The animals that make huge changes are said to have gone through metamorphosis. In this process the full-grown adult looks very different from the newborn creature. After all, tadpoles don't look anything like adult frogs—do they?

This metamorphosis also happens to many different insects! The most famous one is the butterfly, which goes from egg to caterpillar to chrysalis to adult butterfly.

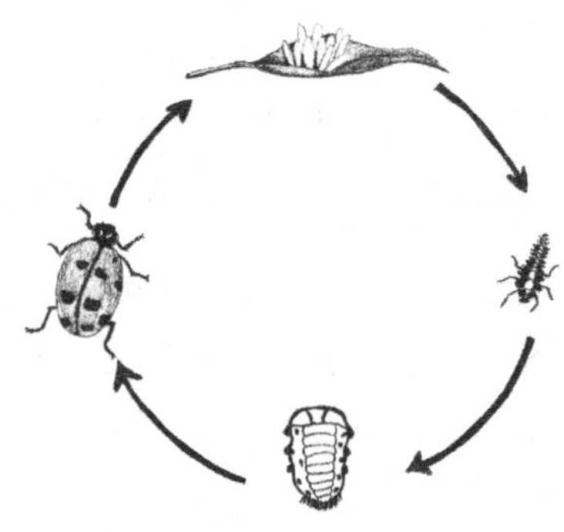

This type of change in insects is called a complete metamorphosis. Other insects, like ladybugs, also undergo complete metamorphosis

On the other hand, insects like the grasshopper undergo a similar type of change with only three stages. Grasshoppers go from egg to nymph to adult grasshopper. This type of change is called an incomplete metamorphosis.

Butterfly Life Cycle Pictures

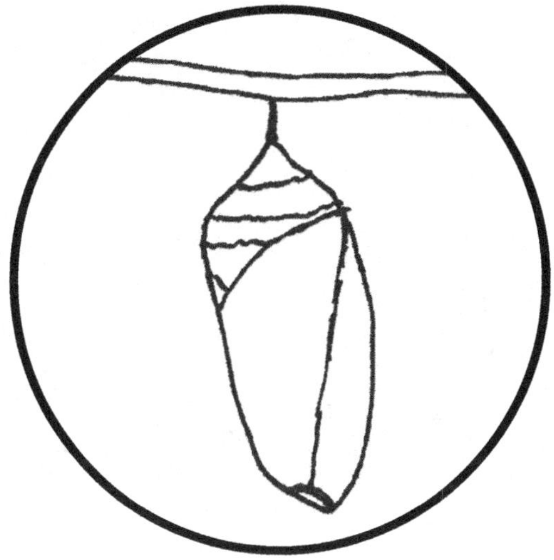

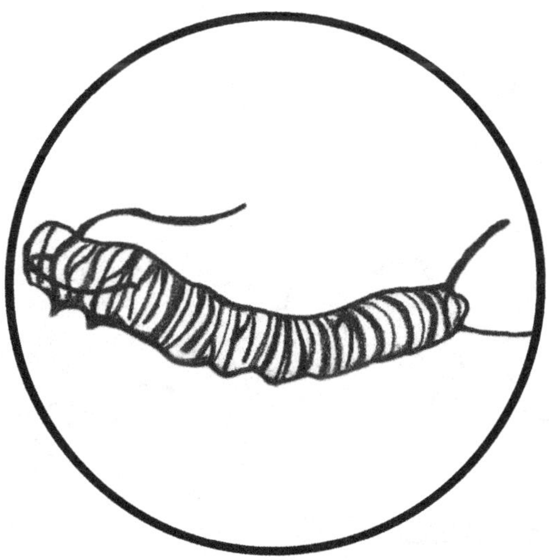

Map of the Earth

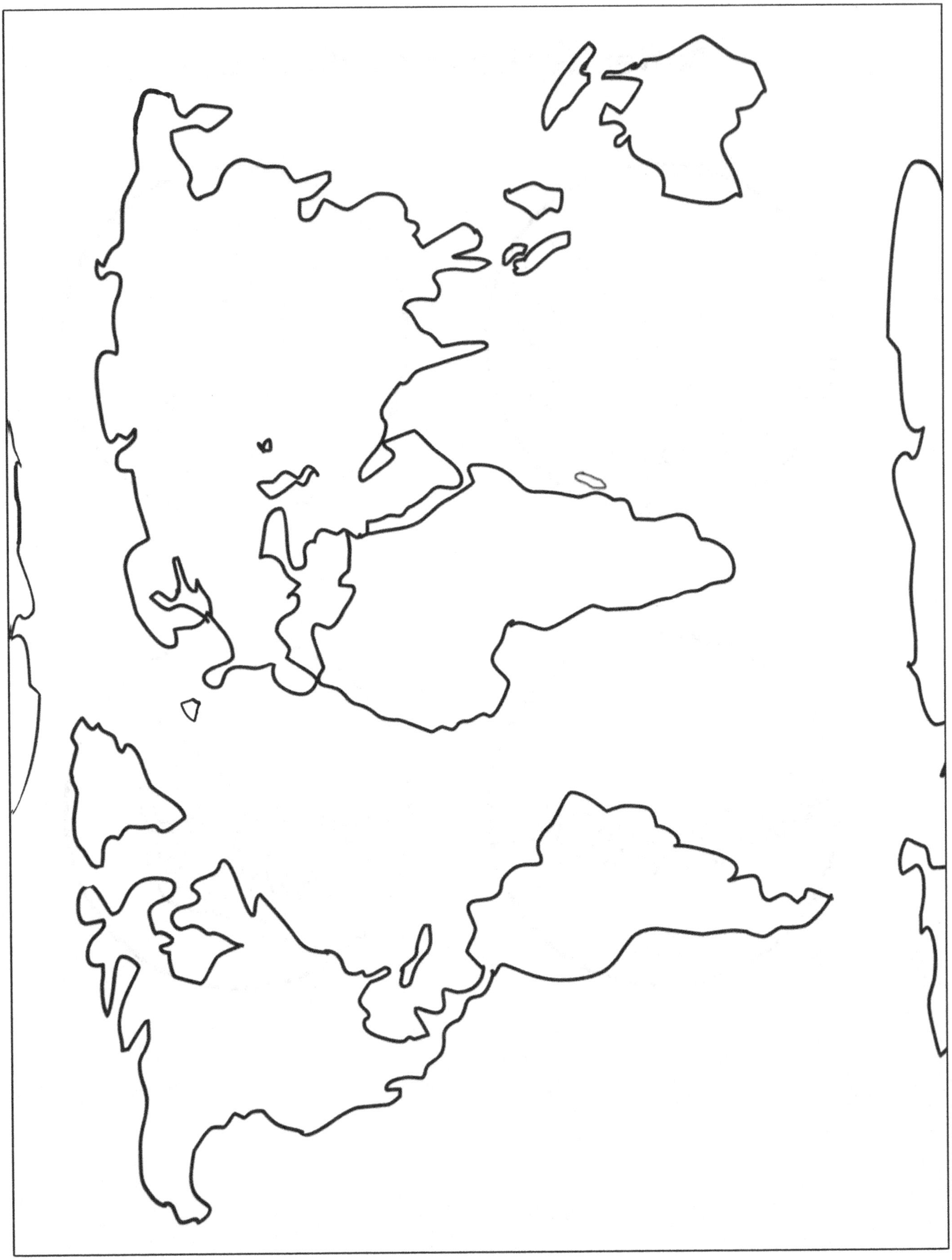

Balance

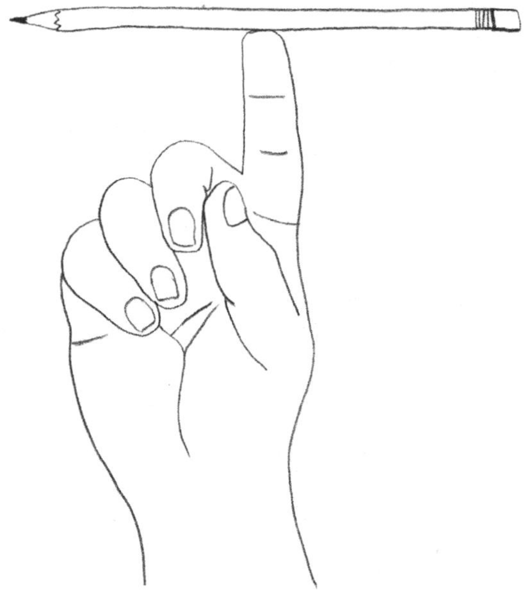

Balance exists when there is an even distribution of weight that allows someone or something to remain upright or steady.

Every object has forces, or pushes and pulls, that act upon it. These forces can cause an object to move in some way or they can cancel each other out, causing the the object to remain at rest.

A book sitting on a table has the force of gravity pulling it down to the floor and the force of the table strength holding it up. It remains where it is, unless we tilt the table, which causes the force of gravity to be greater and the book falls to the floor.

An object is in balance when all the forces that push or pull on it have caused it to remain still.

Christopher Columbus and Balance

There is a popular story about Christopher Columbus and balance. The story says that several of his critics were trying to downplay his discovery of the Americas and to question Columbus's intelligence.

So, Christopher challenged them to make an egg stand on its tip. After much effort, they gave up. Columbus stepped up and tapped the egg on the table, flattening the eggshell on the tip, and the egg was able to stand upright with little effort.

Now, he did crack the egg in the process, but the story says that he silenced his critics by proving that he was the smarter man.

SCIENCE CHUNKS

Glossary

Year C Glossary

A

- Acceleration – A change in the speed or direction of an object.
- Alloy – A mixture of two or more metals or a metal and a non-metal.
- Arthropod – An animal with a jointed body, such as an insect or spider.

B

- Bacteria – Single-celled organisms that live almost everywhere.
- Balance – The point at which all the forces acting on an object cancel each other out.

C

- Centripetal Force – A force that keeps an object moving in a circle.
- Continent – A large area of land on Earth.

D

- Density – A measure of the amount of matter (mass) in a substance compared to its volume.

E

- Element – A substance made up of one type of atom, which cannot be broken down by chemical reaction to form a simpler substance.
- Equator – The most important line of latitude, which divides the Earth in half.
- Essential Element – An element that is essential to life on earth, such as carbon, hydrogen, nitrogen, and oxygen.

F

- Force – A push or a pull on an object.
- Friction – The force that causes moving objects to slow down when they are touching.
- Fungus – A living thing that absorbs food from living or dead matter.

G

- Gravity – The pulling force that attracts objects to each other.

H

I

- Inert – An element that is completely nonreactive.
- Inertia – The tendency of objects to resist a change in their movement.
- Insect – An invertebrate animal that has three body parts (head, thorax, and abdomen) and six legs.
- Invertebrate – An animal without a backbone.
- Ion – An atom or group of atoms that has become charged by gaining or losing one or more electrons.

J

K

L

- Living – Something that eats, breathes, and moves, like you.
- Lubricant – A substance used to reduce friction.

M

- Magma – Molten rock found under earth's surface.
- Magnet – An object that attracts iron, steel, and metals.
- Magnetism – The force that causes magnets and metals to stick together.
- Mass – The amount of matter contained in an object.
- Metal – The largest class of elements, usually they are shiny and solid at room temperature.
- Metalloid – An element that shares some of the properties of metals and nonmetals.
- Metamorphosis – A process by which an animal's body changes shape during its life cycle.
- Mollusk – A group of invertebrate animals, including snails, slugs, and octopuses.

- Momentum – A measure of an object's tendency to continue moving.
- Mountain Range – A row of mountains that form a sort of chain across the land.

N

- Nonliving – Something that is not alive.
- Nonmetal – A class of elements that can be non-shiny solids or gases.

O

- Oxidation – A chemical reaction in which a substance combines with oxygen.

P

- Periodic Table – A systematic arrangement of the elements in order of increasing atomic number.
- Protist – A living thing whose body is made from a single cell.

Q

R

- Radioactive Decay – The process by which a nucleus ejects particles through radiation, becoming the nucleus of a series of different elements until stability is reached.
- Radioactivity – The release of radiation from the nuclei of unstable atoms.
- Reactive – The tendency of a substance to react with other substances.
- Recycling – The act of turning waste, like used cans, bottles, and paper, into new, usable materials.
- Refraction – The bending of light as it passes through a different medium.
- Resultant force – The combined effect of the forces acting on an object.

S

- Shellfish – An aquatic invertebrate animal having a shell.
- Solution – A mixture that consists of a substance dissolved in a liquid.
- Speed – A measure of how fast an object is moving.

- Spore – A microscopic package of cells produced by a fungus or plant that can grow into a new individual.

T

- Torque – The force you add to make something rotate.

U

V

- Velocity – A measure of how fast an object is moving in a given direction.

W

- Weight – A measure of the strength of the pull of gravity on an object.

X

Y

Z

SCIENCE CHUNKS

Review Sheets

Living World Review Sheet

1. Circle the characterisitics of life.

Breathing Moving Growing

Changing Eating

Using energy Having cells Reproducing

2. Circle the living things:

Airplane Giraffe Human Being

Snowflake Flower

3. Which living thing is one of the smallest things alive?

Mouse Virus Elephant

4. Circle the protists:

Algae Cats Protozoans

Amoebas Bacteria

5. All fungus release ________________ to reproduce.

Seeds Spores Captain Crunch

6. Molds (are / are not) found on decaying fruit.

7. **True or False:** A habitat is the natural environment of an animal.

Invertebrates Review Sheet

1. Invertebrates are animals ____________ backbones.

   with without

2. Snails live ________________________.

   only on land only in the water both on land and in water

3. Octopuses have ______ arms, called tentacles.

   four six eight ten

4. Crabs have two front ___________ to help them get food and fight off attackers.

   spears hands pincers

5. **True or False:** Shrimps live in oceans, rivers, and lakes.

6. Spiders have ______ legs.

   four six eight ten

7. Insects have (one three five) body parts and (four six eight) legs.

8. **True or False:** Ants live together in colonies.

9. The series of changes that an animal goes through in its lifetime is called a(n) ______________________.

wheel   life cycle   amazing trip

10. **True or False:** Butterflies turn into caterpillars through metamorphosis.

11. The wings of a beetle are protected by a ____________ shell.

hard   soft

Planet Earth Review Sheet

1. Match the layers of the earth with their description.

   Crust _____

   Mantle _____

   Core _____

2. The day/night cycle is caused by ______________________________.

   The rotation of the Earth   A curtain of clouds that come in and out

3. **True or False:** A continent is a large area of land found on the Earth.

4. A row of mountains is called a mountain __________.

5. Mountains (can / cannot) be formed when two pieces of the Earth's crust push against each other.

6. **True or False:** Maps can tell us information about an area.

7. **True or False:** There is only one kind of map.

8. Circle all the things you can do to help take care of the Earth:

recycle turn off the lights plant trees

pick up trash walk or ride your bike

Periodic Table Review Sheet

1. Fill in the blanks with atomic number, atomic mass, and chemical symbol.

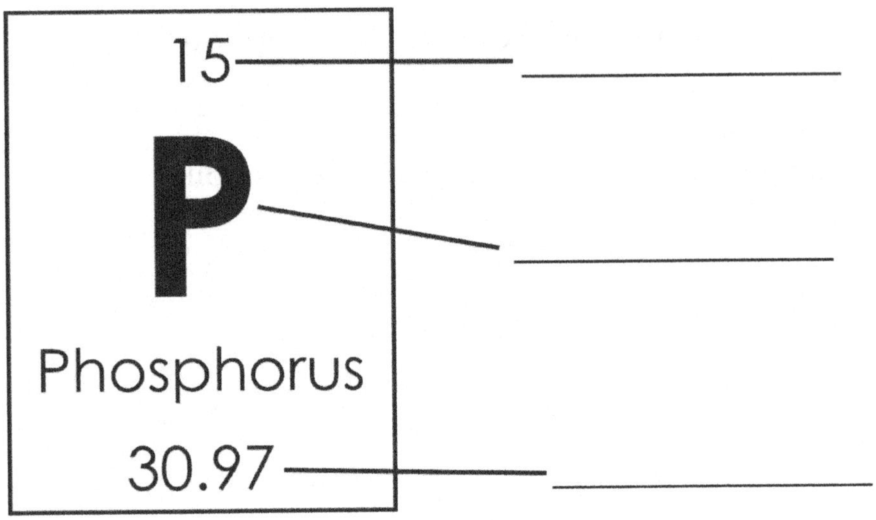

2. **True or False:** The periodic table was first designed by Dmitri Mendeleev.

3. **True or False:** An element is made up of multiple different atoms.

4. **True or False:** The alkali metals are a very reactive group.

5. Hydrogen exists as a ________________ on Earth.

6. Circle all the characteristics of alkaline earth metals.

   soft metals   often bonded to oxygen   react easily

7. **True or False:** Magnesium is a gas that burns with a bright purple-blue light.

8. The amazing ability of transition metals is that they can ___________________ with a variety of other elements to form alloys.

9. **True or False:** Iron can be found in the core of the Earth.

10. **True or False:** Boron is the only element in the boron elements groups that is different than the rest.

11. **True or False:** Boron is the only metal in the group of nonmetal boron elements.

12. Why is carbon so important to us?

____

____

____

13. **True or False:** The nitrogen elements have a mish-mash collection of properties.

14. Nitrogen makes up ___________________ of the air on Earth.

20 %   50 %   80 %

15. The oxygen elements are (not important / important) to life and industry.

16. Why is oxygen so important to us?

____

____

____

17. **True or False:** The elements in the halogens group are not very reactive.

18. The noble gases are the (most / least) reactive elements in the periodic table.

19. **True or False:** Helium is heavier than air.

20. **True or False:** Many of the elements in lanthanide group are naturally occurring.

21. **True or False:** Many of the elements in the actinide group are radioactive.

Curie Review Sheet

1. Match the following words to their definitions.

_____ Magnet	A. The release of radiation from the nuclei of unstable atoms.
_____ Magnetism	B. An object that attracts iron, steel, and metals.
_____ Radioactivity	C. The force that causes magnets and metals to stick together.

2. **True or False:** Marie Curie was always interested in the sciences.

3. **True or False:** Marie Curie found it very easy to isolate radium.

4. **True or False:** Marie Curie discovered both polonium and radium.

5. Marie Curie won _______ Nobel Prizes.

6. Why do you think someone should learn about Marie Curie?

____

____

____

____

____

Force Review Sheet

1. **True or False:** All forces must be visible.

2. A ____________________ force is a force that needs two objects to touch.

3. **True or False:** When the resultant force is equal to zero, an object is balanced.

4. **True or False:** When an object is at rest, there are no forces acting on it.

5. An object's ______________________________ is the point at which the whole weight of an object seems to act.

6. **True or False:** Weight is the amount of matter in an object, while mass is the measure of the strength of the pull of gravity on an object.

7. **True or False:** A lubricant helps to increase friction.

8. ______________________ is a type of friction that occurs between air and an object moving through it.

9. An object floats because it weighs ____________________ than the fluid it moves out of the way.

10. **True or False:** Archimedes's principle says that the upthrust acting on an object is equal to the weight of the fluid that the object displaces.

Motion Review Sheet

1. (Inertia / Momentum) is the tendency of objects to resist a change in their movement. (Inertia / Momentum) is a measure of an object's tendency to continue moving.

2. **True or False:** An object in motion stays in motion unless acted upon by an outside force.

3. Newton's 2nd law of motion states that the (greater / lesser) the force on an object, the (greater / lesser) the change in motion.

4. Scientists use an object's ____________________ to describe its motion.

   speed velocity acceleration all of the above

5. **True or False:** Acceleration is a change in an object's speed or direction.

6. **True or False:** Terminal velocity is the maximum velocity reached by a falling object.

7. **True or False:** Objects moving in a circle do not travel in a straight line.

Made in the USA
Monee, IL
12 May 2026

49985522R00072